Basic Principles of

COLLOID SCIENCE

Royal Society of Chemistry Paperbacks

Royal Society of Chemistry Paperbacks are a series of inexpensive texts suitable for teachers and students and giving a clear, readable introduction to selected topics in chemistry. They should also appeal to the general chemist. For further information on selected titles contact:

Sales and Promotion Department
The Royal Society of Chemistry
Burlington House
Piccadilly
London W1V 0BN

Titles Available

Water *by Felix Franks*
Food – The Chemistry of Its Components *by T. P. Coultate*
Analysis – What Analytical Chemists Do *by Julian Tyson*
Basic Principles of Colloid Science *by D. H. Everett*

How to Obtain RSC Paperbacks

Existing titles may be obtained from the address below. Future titles may be obtained immediately on publication by placing a standing order for RSC Paperbacks. All orders should be addressed to:

The Royal Society of Chemistry
Distribution Centre
Blackhorse Road
Letchworth
Herts. SG6 1HN

Telephone: Letchworth (0462) 672555
Telex: 825372

Royal Society of Chemistry Paperbacks

Basic Principles of
COLLOID SCIENCE

D. H. EVERETT, F.R.S.
Department of Physical Chemistry
University of Bristol

British Library Cataloguing in Publication Data
Everett, D. H. (Douglas Hugh), 1916–
Basic principles of colloid science.
1. Colloids
I. Title
541.3′451

ISBN 0-85186-443-0

Published by the Royal Society of Chemistry
Burlington House, Piccadilly, London W1V 0BN

Typeset by KEYTEC, Bridport, Dorset
Printed by Whitstable Litho Ltd., Whitstable, Kent

Preface

Colloid science is experiencing a renaissance. The beginnings of this new phase can be traced back some fifty years when a scientific understanding of at least some colloidal phenomena began to evolve. Since then activity has increased steadily. Fundamental knowledge has developed rapidly and the resulting insights have been exploited extensively in industry. Many empiricisms which for generations have guided practical applications have gradually been shown to have their origin in the laws of physics and chemistry. There is still much to be learned, but a stage is being reached at which it is becoming possible to present a general account of the main themes of colloid science in terms of basic physical chemistry. That is one of the objectives of the present book.

It has also become clear that there is a need for a book that presents an outline of colloid science starting from a relatively elementary knowledge of science. Several such books have appeared in the past but all are now out of print, and in any case they are somewhat outdated in view of recent developments. The present book will, it is hoped, fill this gap and both provide an introduction to colloids for those with a basic familiarity with physical chemistry and serve as a jumping-off point for those wishing to go more deeply into the many fascinating areas which constitute the broad range of fundamental and applied colloid science.

One of the major features of the development of colloid science has been the impressive advances that have been made in colloid technology. However, although many large companies have exploited recent developments, it is unfortunately the case that smaller companies have often failed to recognise the potential applications of colloid science in enhancing the efficiency of their processes or the quality and range of their products. Attention is therefore drawn to some of those areas of the subject that have been, or can be, of industrial importance. In a book of this size it is impossible to deal in detail with such matters, but Chapter 14 outlines some of the industrial applications of colloid science. Thus the book as a whole may prove of value both to those entering industry with little previous knowledge of colloids and to those already in industry wishing to become familiar with recent ideas.

The structure of the book is such that the earlier chapters can be read by students studying chemistry at A-level (in conjunction, for example, with the Nuffield Advanced Special Option on Surface Chemistry). Some simple experiments on colloidal systems described in Appendices I and II are suitable for use at this level or in undergraduate courses. Later it is necessary to employ a more sophisticated mathematical approach, but this is kept to a minimum and should not deter those with a modest mathematical background. Some of the more detailed theoretical topics are relegated to Appendices III—VI.

It will be seen that the main emphasis of the book is on disperse systems. While the crucial importance of surface chemistry is stressed, it has been thought preferable to omit discussion of the adsorption of gases and vapours by solids and to leave out some of the very important phenomena associated with capillarity, wetting, and spreading.

A book of this kind can only expect to provide one perspective of an immense subject. Many readers will, one hopes, wish to broaden their knowledge and understanding, and to this end Appendix VII lists references for further reading. It includes: relatively 'popular' accounts in, for example, the *Scientific American*; papers aimed mainly at the educationalist in the *Journal of Chemical Education*; and those with industrial implications in *Chemistry and Industry*. In addition, some of the books and articles deal in a more advanced fashion with recent developments which take the reader up to the frontiers of current research.

In writing this book I have been influenced by many colleagues and friends. In particular, I wish to acknowledge my gratitude to those who have read and commented on either the whole or parts of the manuscript. These include Dr. A. Couper, Dr. T. H. K. Barron, Dr. B. Vincent, Dr. J. W. Goodwin, Dr. S. Lubetkin, and Dr. A. L. Smith. Their help and advice have been greatly appreciated. I am also indebted to Mr. S. R. Neck, who has over the years given me invaluable help in devising and constructing demonstration models and experiments to illustrate many of the topics dealt with in this book.

This book would not have been written but for the development of the strong School of Colloid Science which has been built up in Bristol during the past 25 years. I am grateful, therefore, to all those, staff and students, who have contributed to the success of this venture, and in particular to Professor Ron Ottewill, F.R.S., for the central role he has played throughout this period.

Finally, I owe an immense debt of gratitude to my wife for her unfailing support and encouragement throughout the writing of this book.

D. H. Everett
Bristol, October 1987

Contents

Chapter 1

What are Colloids?

To some the word 'colloidal' conjures up visions of things indefinite in shape, indefinite in chemical composition and physical properties, fickle in chemical deportment, things infilterable and generally unmanageable.

Hedges, 1931

INTRODUCTION

The above remarks reflect the impression created by many textbooks of physical chemistry – if they deign to mention colloids at all. In fact, in both its experimental and theoretical aspects, and no less important in its technological applications and in the appreciation of its biological implications, colloid science has made impressive progress in the last few decades. In the following chapters an attempt is made to summarise the basic concepts of colloid science and to dispel some of the doubts expressed in the above quotation.

A full understanding of the properties of colloids calls upon a wide range of physical and chemical ideas, while the multitude of colloidal systems presented to us in nature, and familiar in modern society, exhibit a daunting complexity. It is this that has delayed the development of colloid science, since a detailed and fundamental theoretical understanding of colloidal behaviour is possible only through a thorough knowledge of broad areas of physics, chemistry, and mathematical physics, together in many instances with an understanding of biological structures and processes. On the experimental side there is an ever-increasing emphasis on the application of modern physical techniques to colloidal problems. Colloid science is thus a truly interdisciplinary subject.

Nevertheless, despite the sophistication needed for the development of a complete quantitative theory of colloids, the basic principles that underlie many colloid problems can be seen as extensions to such systems of the fundamental concepts of physical chemistry. One important objective of this book is to emphasise the close link between colloid science and physical chemistry and to show how a broad understanding can be built up on a few relatively simple physico-chemical ideas. We shall not only seek common features revealed by experimental study but also, of much greater significance, try to identify the fundamental concepts that link together many apparently unconnected aspects of the subject.

DEFINITION OF COLLOIDS

In setting out to define the scope of colloid science, it should first be said that any attempt to lay down too rigid a scheme of definitions and nomenclature is likely to be unnecessarily restrictive. Rather than try at the outset to develop a formal definition, it is preferable to describe examples of systems to which the term 'colloidal' is now applied.*

An essential part of any study of physics and chemistry involves first the recognition of three states of matter – solid, liquid, and gas – and a general discussion of the transformations – melting, sublimation, and evaporation – between them. Pure substances are considered, and then attention passes to solutions which are homogeneous mixtures of chemical species dispersed on a molecular scale. What remained largely unrecognised until about a century and a half ago was that there is an intermediate class of materials lying between bulk and molecularly dispersed systems, in which, although one component is finely dispersed in another, the degree of subdivision does not approach that in simple molecular mixtures. Systems of this kind, *colloids*, have special properties which are of great practical importance, and they were appropriately described by Ostwald as lying in the World of Neglected Dimensions. They consist of a *dispersed phase* (or *discontinuous phase*) distributed uniformly in a finely divided state in a *dispersion medium* (or *continuous phase*).

As familiar examples of colloidal systems we cite the following: fogs, mists, and smokes (dispersions of fine liquid droplets or solid particles in a gas – *aerosols*); milk (a dispersion of fine droplets of

* The etymology of the term colloidal (glue-like) introduced by Thomas Graham is now largely irrelevant.

fat in an aqueous phase – *emulsions*); paints, muds, and slurries (dispersions of fine solid particles in a liquid medium – *sols* or *colloidal suspensions*); jellies (dispersions of macromolecules in liquid – *gels*); opal and ruby stained glass (dispersions, respectively, of solid silica particles in a solid matrix or of gold particles in glass – *solid dispersions*). So-called (and miscalled) *photographic emulsions* are dispersions of finely divided silver halide crystallites in a gel – in a sense they are a colloid within a colloid. In *association colloids* molecules of soap or other surface-active substances are associated together to form small aggregates (*micelles*) in water. The aggregates formed by certain substances may adopt an ordered structure and form *liquid crystals*. Many biological structures are colloidal in nature. For example, blood is a dispersion of corpuscles in serum, and bone is essentially a dispersion of a calcium phosphate embedded in collagen.

In the above examples, which may be called *simple colloids*, a clear distinction can be made between the disperse phase and the dispersion medium. However, in *network colloids* this is hardly possible since both phases consist of interpenetrating networks, the elements of each being of colloidal dimensions. Porous solids, in which gas and solid networks interpenetrate, two-phase glasses (opal glasses), and many gels are examples of this category.

Furthermore, there are other instances (*multiple colloids*) that may involve the co-existence of three phases of which two (and sometimes three) phases are finely divided. One example is a porous solid partially filled with condensed vapour, when both the liquid and vapour phases within the pores are present in a finely divided form; a similar situation arises when oil and water co-exist in the pores of an oil-bearing rock, also in frost heaving when water and ice co-exist in a porous medium. *Multiple emulsions* consist for example of finely divided droplets of an aqueous phase contained within oil droplets, which themselves are dispersed in an aqueous medium.

Some of the more important types of colloidal systems outlined above are summarised in Table 1.1. For simplicity we shall limit ourselves in this book to a discussion of simple colloids, although the ideas developed can be extended and applied to more complex systems.

The fundamental question which has to be answered is 'What do we mean by "finely divided"?' It turns out, for reasons which will soon be apparent, that systems usually exhibit properties of a specifically 'colloidal character' (which we shall explain in more

Table 1.1 *Some typical colloidal systems.*

Examples	*Class*	*Nature of the disperse phase*	*Nature of the dispersion medium*
Disperse systems			
Fog, mist, tobacco smoke, 'aerosol' sprays	Liquid aerosol or aerosol of liquid particles [a]	Liquid	Gas
Industrial smokes	Solid aerosol or aerosol of solid particles [a]	Solid	Gas
Milk, butter, mayonnaise, asphalt, pharmaceutical creams	Emulsions	Liquid	Liquid
Inorganic colloids (gold, silver iodide, sulphur, metallic hydroxides, *etc.*), paints [b]	Sols or colloidal suspensions	Solid	Liquid
Clay slurries, toothpaste, muds, polymer latices	When very concentrated called a paste	Solid	Liquid
Opal, pearl, stained glass, pigmented plastics	Solid suspension or dispersion	Solid	Solid
Froths, foams	Foam [c]	Gas	Liquid
Meerschaum, expanded plastics	Solid foam	Gas	Solid
Microporous oxides, 'silica gel', porous glass, microporous carbons, zeolites	Xerogels[d]		
Macromolecular colloids			
Jellies, glue	Gels	Macro-molecules	Solvent
Association colloids			
Soap/water, detergent/water, dye solutions	—	Micelles	Solvent

Table 1.1 *(cont.)*

Examples	*Class*	*Nature of the disperse phase*	*Nature of the dispersion medium*
Biocolloids			
Blood		Corpuscles	Serum
Bone		Hydroxy-apatite	Collagen
Muscle, cell membranes		Protein structures, thin films of lethecin, *etc.*	
Three-phase colloidal systems (multiple colloids)			
		Coexisting phases	
Oil-bearing rock	Porous rock	Oil	Water
Capillary condensed vapours	Porous solid	Liquid	Vapour
Frost heaving	Porous rock or soil	Ice	Water
Mineral flotation	Mineral	Water	Air bubbles or oil droplets
Double emulsions	Oil	Aqueous phase	Water

[a] Preferred nomenclature according to IUPAC recommendations. [b] Many modern paints are more complex, containing both dispersed pigment and emulsion droplets. [c] In a foam it is usually the thickness of the film of dispersion medium which is of colloidal dimensions, although the dispersed phase may also be finely divided. [d] In some cases both phases are continuous, forming interpenetrating networks both of which have colloidal dimensions.

detail in later chapters) when the dimensions of the dispersed phase lie in the range 1—1000 nm, *i.e.* between 10 Å and 1 μm.*

These limits are not rigid, for in some special cases (*e.g.* emulsions and some slurries) particles of larger size are present. Moreover, it is not necessary for all three dimensions to lie below 1 μm, since colloidal behaviour is observed in systems containing fibres in which only two dimensions are in the colloid range. In other systems, such as clays and thin films, only one dimension is in the colloid range. This is illustrated schematically in Figures 1.1 and 1.2, while Figure 1.3 shows electron microscope photographs of colloidal particles of several types.

* These dimensions are below the limits of resolution of simple optical microscopes so that direct imaging and measurement of the sizes of colloidal particles only became possible with the development of electron microscopes.

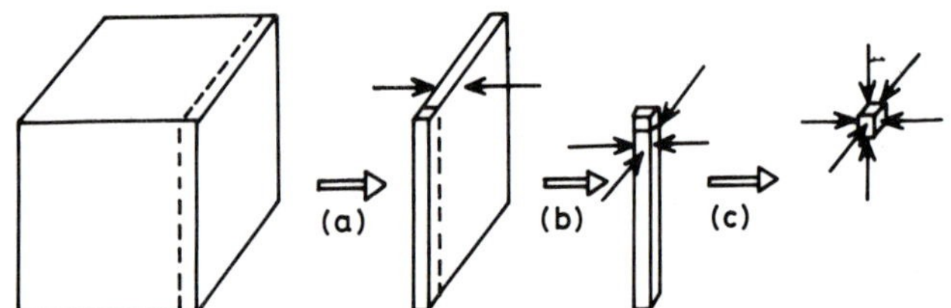

Figure 1.1 *Schematic representation of the subdivision of a cube to give colloidal systems of different kinds:* (a) *slicing of a cube leads to a* laminated disperse system *with one dimension in the colloid range,* (b) *cutting a sheet into narrow strips leads to* fibrillar disperse systems *with two dimensions in the colloid range,* (c) *cutting of rods or fibrils into particles leads to* corpuscular disperse systems *with all three dimensions in the colloid range.*

(Adapted from A. von Buzagh, 'Colloid Systems', Technical Press, London, 1937)

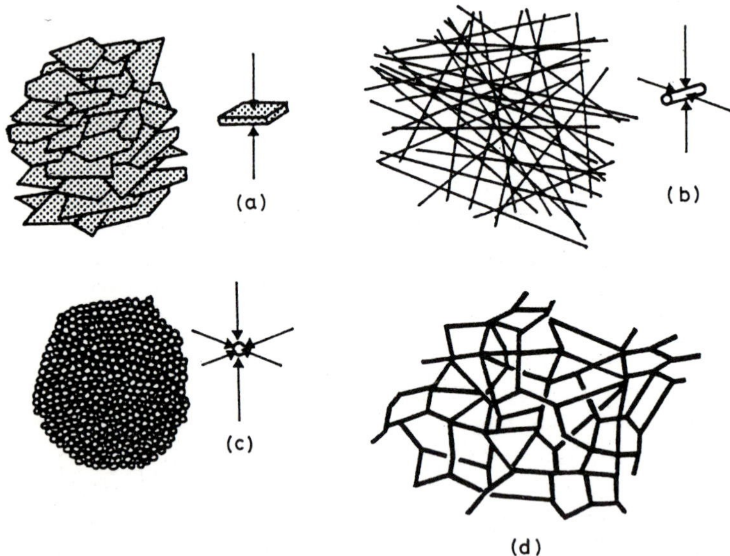

Figure 1.2 *Schematic representation of colloid systems of various types:* (a) *kaolinite,* (b) *Plaster of Paris, cement, asbestos,* (c) *polymer latices,* (d) *network structures, e.g. porous glass, gels.*

Colloids in which the particle size is below about 10 nm often require special consideration. One example of such particles are the nuclei which initiate bulk phase changes, while the justification for including macromolecular solutions and association colloids within this classification arises from the fact that the particles within them are either macromolecules of considerable length, which even when coiled up have diameters of well over 1 nm, or aggregates of smaller molecules forming micelles of a size falling within the colloid size range. Biocolloids again have their individual characteristics, but once more the presence of structures of colloidal dimensions justifies their inclusion as examples of colloids. The limit below which colloid behaviour merges into that of molecular solutions is usually presumed to be around 1 nm (10 Å).

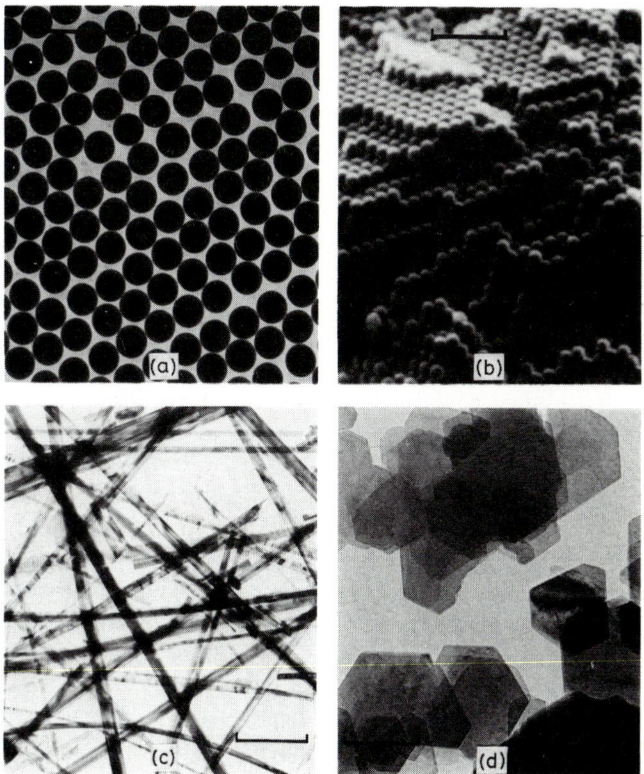

Figure 1.3 *Electron micrographs of colloidal materials in which three, two, and one dimensions lie in the colloid range (bars indicate 1 μm):* (a) *spherical particles of monodisperse polystyrene latex,* (b) *packed spherical particles of polystyrene latex,* (c) *fibres of chrysotile asbestos,* (d) *thin plates of kaolinite.*

(By Dr. D. W. Thompson, School of Chemistry, University of Bristol)

An alternative subdivision of colloids which has been widely used in the past is into *lyophobic* (or *hydrophobic*, if the dispersion medium is water) and *lyophilic* (*hydrophilic*, in water) colloids, depending on whether the particles can be described in the former case as 'solvent hating' or in the latter case as 'solvent loving'. These characteristics are deduced from the conditions required to produce these colloids and from the means available for their redispersion after flocculation or coagulation. It will become apparent later that, while this subdivision has many useful aspects, it is neither entirely logical nor sufficiently all-embracing, and we shall make only limited use of it.

COLLOIDS AND SURFACE CHEMISTRY

Because of the range of dimensions involved in colloidal structures, the surface-to-volume ratio is high and a significant proportion of the molecules in such systems lie within or close to the region of inhomogeneity associated with particle/medium interfaces. These molecules will have properties (*e.g.* energy, molecular conformation) different from those in the bulk phases more distant from the interface. It is then no longer possible (as we do in bulk thermodynamics) to describe the whole system simply in terms of the sum of the contributions from the molecules in the bulk phases, calculated as though both phases had the same properties as they have in the bulk state. A significant and often dominating contribution comes from the molecules in the interfacial region. This is why surface chemistry plays such an important part in colloid science and why colloidal properties begin to become evident when the particle size falls below 1 μm. We can see this in the following way.

The surface area associated with a given mass of material subdivided into equal-size particles increases in inverse proportion to the linear dimensions of the particles. Thus the area exposed by unit mass (the *specific surface area*, a_s) is given by $6/\rho d$, where ρ is the density of the material and d is the edge length in the case of cubic particles or the diameter in the case of spheres. If the material is made up of molecules of linear dimension h and molecular volume $\sim h^3$, then the fraction of molecules in the surface layer is given approximately by $6(h/d)$. Thus for a substance of molar volume 30 $cm^3\,mol^{-1}$ or of molecular volume 0.05 nm^3 (*e.g.* silver bromide) $h = 0.37$ nm. For a 1 cm cube only two or three molecules in ten million are surface molecules, and these have a negligible influence on its properties. However, when divided into 10^{12} particles of 1 μm, one molecule in four hundred and fifty is a surface molecule, and the properties of the system begin to be affected. At 10 nm the ratio rises to nearly one in four and surface effects dominate. Beyond this it is hardly possible to decide what we mean by a surface molecule, and, as indicated above, special considerations apply to the size range 1—10 nm. To illustrate this point, Figure 1.4 shows the variation of the percentage of surface molecules with particle size for the typical case of silver bromide.

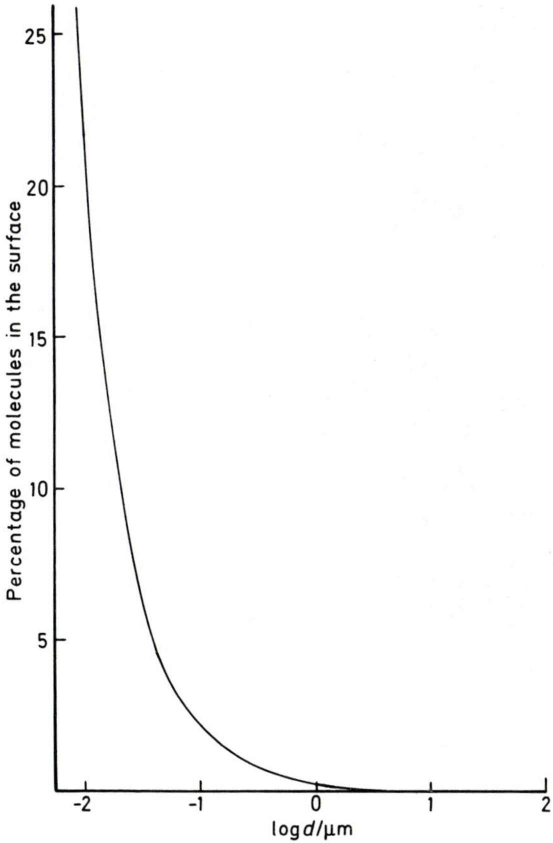

Figure 1.4 *Variation of the percentage of molecules in the surface as a function of particle size for a substance with a molar volume of 30 cm^3 mol^{-1}.*

This approach to colloids, emphasising the importance of surface or interfacial properties, suggests a more meaningful description of colloids as *microheterogeneous systems*, the microheterogeneity being characterised by lengths in the range 1—1000 nm.

It should be noted, however, that some typically colloidal phenomena, such as light scattering, are exhibited (though very weakly) by systems in which the microheterogeneity arises from random kinetic fluctuations in density in an otherwise uniform system of small molecules such as a gas or a liquid, while in some cases (*e.g.* suspensions of relatively coarse solid particles) certain colloid-like properties may persist to particle sizes much larger than the above maximum.

NOMENCLATURE*

Before proceeding further, it will be helpful to introduce a number of additional terms which are widely used in the description of colloidal behaviour.

Disperse systems in which all the particles are of (approximately) the same size are said to be *monodisperse* (or *isodisperse*); conversely, if a range of particle sizes is present, they are *polydisperse*. In certain circumstances, to be discussed in greater detail later, the particles of a dispersion may adhere to one another and form aggregates of successively increasing size which may, despite the tendency of thermal motion to keep them in suspension, separate out under the influence of gravity. The nature of the aggregated material may depend on the conditions of its formation, or it may change with time. An initially formed, rather open aggregate is called a *floc* and the process of its formation *flocculation*. The floc may or may not separate out. If the aggregate changes to, or is produced in, a much denser form, it is said to undergo *coagulation* with the formation of a *coagulum*. An aggregate usually separates out either by *sedimentation* (if it is more dense than the medium) or by *creaming* (if it is less dense than the medium). Since in many cases it is not readily apparent

* Reference should be made to the recommendations of the International Union of Pure and Applied Chemistry (IUPAC) 'Definitions, Terminology and Symbols in Colloid and Surface Chemistry', *Pure and Applied Chemistry*, 1972, **31**, 579—638, for a fuller discussion. More specific IUPAC recommendations on particular aspects of colloids are to be found in *Pure and Applied Chemistry*, 1974—86; see Appendix VII.

which type of aggregate is formed, the terms flocculation and coagulation have often been used interchangeably, but the more specific meanings introduced above are gaining more general acceptance. One characteristic which is sometimes used to infer a distinction between the two is whether aggregation is reversible. It is usually supposed that coagulation is irreversible whereas flocculation can be reversed in the process of *deflocculation*.* However, conditions can sometimes be found under which even coagulated systems can be redispersed.

The meaning of the term *stability* as applied to colloid systems is discussed in Chapter 2.

AN HISTORICAL PERSPECTIVE

Although the true nature of colloids was not appreciated until relatively recently, Man has observed and made use of colloidal systems and their properties since the earliest days of civilisation. Moreover, colloids have played in geological time, and play today, an important role in many natural phenomena. Perhaps the oldest record of a colloidal phenomenon is that of the deposition of silt at river mouths mentioned in the Babylonian Creation myth – which, incidentally, was inscribed on tablets of clay, themselves an example of a colloidal material – while the Book of Genesis refers to clouds and the fall of rain. But early Man must also have been familiar with many other colloidal phenomena, such as the effect of walking on wet sand and the treachery of quicksands. He soon exploited them in the preparation of butter, cheese, and yoghurt and in the making of bread. His early technology, too, often depended on colloids and their properties: the making of bricks, the extraction of glue from bones, and the preparation of inks and pigments are a few examples. Indeed, there can have been few aspects of his domestic life that were independent of the behaviour of colloids, either of natural occurrence or prepared by him. The same is even more true today: the list of colloids and colloidal processes of vital importance to modern living and industrial technology is almost unlimited. Their diversity can be appreciated by quoting just a few examples, which include the following: the products of

* Beware that occasionally the term deflocculation is used, misleadingly, to denote the process floc $\rightarrow$ coagulum.

modern food technology; pharmaceutical and cosmetic preparations of many kinds; agricultural and horticultural chemicals; paints, dyestuffs, and paper; the processes involved in mineral extraction, oil recovery, water treatment, photography, biotechnology, and so on. The difference today is that colloid technology is rapidly becoming more rational and scientifically based and is leaving behind many of the empiricisms that characterised those earlier crafts which depended on controlling and using colloidal materials. We shall return to discuss some of these aspects in more detail in Chapter 14.

Despite this long history, the scientific study of colloids is a relatively recent development. It is true that the alchemists prepared and used two important forms of colloidal gold, namely potable gold (supposedly the Elixir of Life) and Purple of Cassius, used to make ruby glass. And Macquer in his Dictionary of Chemistry (1774) speculated that in these gold was present in a finely divided form. But the first experimental studies date from the early years of the nineteenth century, when Selmi (1845) prepared what were then called demulsions of sulphur and silver halides. It was not until 1856 that Michael Faraday made the first systematic study of colloidal gold, which will be outlined in the following section, and put forward ideas which can still be seen in modern theories concerning the factors responsible for the stability of these dispersions. The word 'colloid' was coined later by Thomas Graham in 1861 to describe systems which exhibited slow rates of diffusion through a porous membrane, of which glue solutions were a typical example. The usefulness of this definition depends on the significant decrease in the diffusion rate when the size of the diffusing particle exceeds a few nanometres. However, such solutions are but one example of the wide class of disperse systems described above, although this was not appreciated immediately.

The slow progress in the understanding of colloidal behaviour compared with that in other branches of chemistry and physics was in large part due to the extreme difficulty of preparing well characterised materials with reproducible properties, as exemplified by the quotation heading this chapter, and in part to the absence of adequate theoretical knowledge to provide a basis for understanding the factors controlling these properties.

Recent progress has followed the reduction and, in some cases, the elimination of these barriers. Thus methods of preparing well characterised colloids have made it possible to perform quantitative and reproducible experiments, while the development of theories of intermolecular forces, electrolyte solutions, and polymers was essential before the concepts they introduced could be brought together and applied in colloid science. Coupled with these factors, and playing an increasingly important role, has been the application to colloids of sophisticated modern instrumental techniques, including high-resolution and scanning electron microscopy, laser light scattering in its various forms, neutron scattering, nuclear magnetic resonance, optical spectroscopy (infrared and Raman in particular), and greatly improved rheological techniques (see Chapter 15).

The guidance provided by basic research into the fundamental factors controlling colloidal behaviour has proved, and is increasingly proving, of immense value in enhancing industrial processes that involve colloids and in developing new processes and products. Among the many examples – some of which will be discussed in Chapter 14 – are the dramatic improvements in paint technology and drug delivery systems which have taken place in the last twenty to thirty years, based in large measure on an increased understanding of the principles of colloid and surface chemistry.

AN ILLUSTRATIVE EXAMPLE: COLLOIDAL GOLD

In later chapters we shall discuss the properties of a variety of colloidal systems. However, as a preparation for our consideration of the general properties and stability of dispersions, it will be useful to use one simple example to outline some of their more important characteristics. It so happens that one of the first colloidal dispersions to have been examined systematically will suit our purpose admirably.

During 1856–7 Michael Faraday first prepared colloidal gold by reducing an aqueous solution of gold chloride with phosphorus to yield a ruby-coloured liquid.* He showed by chemical tests that the gold was no longer present in an ionic form but that reagents that dissolve metallic gold were able to remove the colour. He concluded that the gold was dispersed in the liquid in

* In Appendix I instructions are given for preparing this and some other typical colloidal dispersions; Appendix II outlines some simple experiments to illustrate their properties.

a very finely divided form, the presence of which could be detected by the blueish opalescence observed when a narrow, intense beam of light is passed through the liquid.* He observed that the addition of a small amount of various salts changed the colour from ruby towards blue and that the blue liquid tended to deposit solid. Neither the blue liquid nor these deposits could be changed back to ruby. He found that the gold sol could also be produced in the presence of a warm gelatine solution, which on cooling set to a jelly. Moreover, when prepared in this way addition of salt to the warm solution did not change the colour to blue.

Faraday concluded that the change from ruby to blue resulted from an increase in particle size. Of the particles in the ruby liquid he said, 'Whether the particles be considered as mutually repulsive, or else as molecules** of gold with associated envelopes of water, they differ from those particles which by the application of salt or other substances are rendered mutually adhesive, and so fall and clot together.'

His observations on the jellied samples implied, he believed, 'a like association (of the gold particles) with that animal substance' which explained their stability in the ruby form.

In this series of experiments Faraday thus demonstrated some of the more important properties of colloidal dispersions: light scattering, sedimentation, coagulation by salts, and their 'protection' from the effects of salt by gelatine. His interpretation of these observations was remarkably perceptive, in contrast to the speculations of some of his contemporaries. He correctly surmised that the change induced by changing conditions 'is not a change of the gold as gold, but rather a change in the relations of the surface of the particle to the surrounding medium'.

It is perhaps surprising that Faraday did not examine the effect of an electric current on his gold sols. Had he done so, he would have discovered the one additional factor which in due course provided the clue to many of their properties, namely that colloidal particles in an aqueous medium (except under special circumstances) move under the influence of an electric field. The

* This phenomenon was subsequently investigated by Tyndall and is known as the *Tyndall effect*.

** Note that the modern use of the word 'molecule' was not introduced until Cannizarro publicised the work of Avogadro in 1859.

phenomenon of *electrophoresis* (of which more in Chapter 6) shows that colloidal particles usually carry an electric charge, which in the case of gold sols is negative. We shall have occasion later to discuss the origin of these charges and the factors that determine their sign. For the moment it is sufficient to know that they exist.

With this brief account of the properties of one typical colloidal dispersion we are now in a position to examine in the next two chapters the factors that are responsible for the stability of dispersions.

Chapter 2

Why are Colloidal Dispersions Stable? I Basic Principles

INTRODUCTION

In this and the following chapter we shall be concerned mainly with dispersions of colloidal solid particles in liquids. Two closely related questions arise. First, under what conditions will the dispersion remain in the dispersed state? We need to know the answer to this before suitable methods of preparation of a stable dispersion can be defined and understood.

Secondly, under what conditions will the dispersion flocculate or coagulate? The answer to this is of vital importance in the many practical situations in which colloids must either be avoided or eliminated (*e.g.* in the filtration of precipitates or in water purification).

To understand the nature of these questions, we must first say something about the principles of physico-chemical equilibrium and show how they can be applied to colloidal systems.

THE MEANING OF 'STABILITY'

We have referred to colloidal systems as being either stable or unstable. It is important to be clear about what is implied by these terms and how their usage in colloid science is related to that in other areas of physical chemistry.

It is a fundamental principle of thermodynamics that, if a system is kept at a constant temperature, it will tend to change spontaneously in a direction which will lower its free energy. This is exemplified by the simple mechanical case of a weight that falls under the influence of gravity or of a ball bearing, released from the edge of a saucer, that runs down and settles at the bottom of the saucer. In each case the reverse of the spontaneous process – raising the weight or rolling the ball to the rim of the saucer – is one in which work has to be done on the system.

It is essential to stress that systems only *tend* to transform to states of lower free energy – the change actually occurs only if a suitable mechanism exists which enables it to take place. Thus a weight resting on a table can manifest its tendency to fall only if it is moved to the edge of the table and allowed to drop off.*

A more meaningful analogy for our present purposes is that of a skittle pin on a horizontal platform (Figure 2.1). The free energy of the pin (which in this simple case can be identified with its gravitational potential energy) may be measured, relative to the surface of the table, by the product $mg\Delta h$, where m is the mass of the skittle, g the acceleration due to gravity, and Δh the height of the centre of gravity of the pin above the table. In the configuration shown in Figure 2.1(a) the skittle has a higher free energy than that in the flat position [Figure 2.1(c)]. The tendency to fall to the position of lower free energy cannot, however, manifest itself unless the pin is sufficiently disturbed (*e.g.* by the impact of a ball) so that it reaches the intermediate configuration shown in Figure 2.1(b). The way in which the free energy of the pin varies with the angle of rotation θ is shown in Figure 2.2.

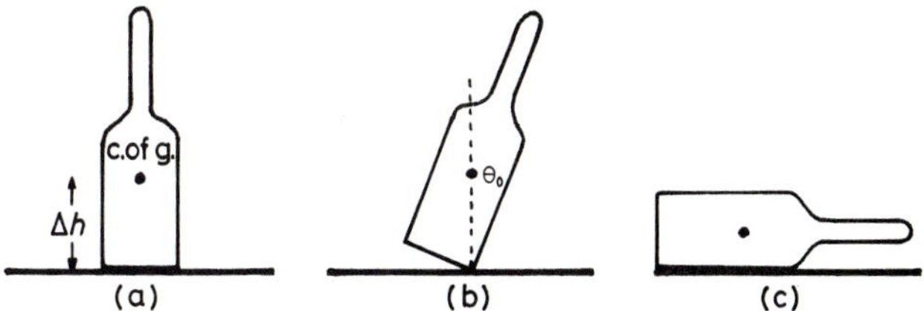

Figure 2.1 *Illustration of three types of equilibrium of a skittle:* (a) *metastable,* (b) *unstable,* (c) *stable.*

* In a similar way, a gaseous mixture of oxygen and hydrogen – which has a much higher free energy than the corresponding amount of liquid water – is unable to manifest this tendency to chemical reaction unless it is sufficiently disturbed (*e.g.* by an electric spark) to enable the molecular processes of chemical reaction to occur.

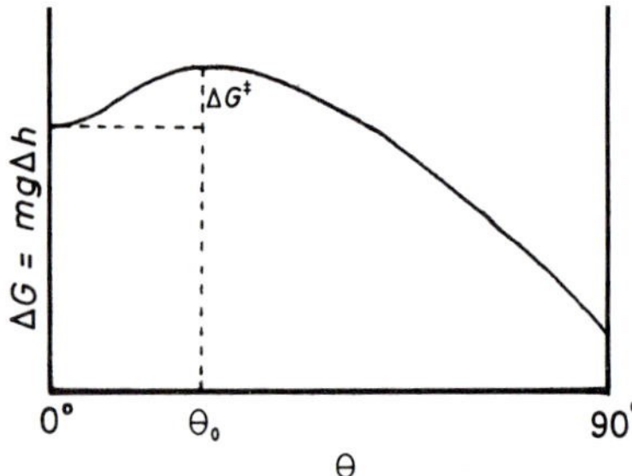

Figure 2.2 *Variation of gravitational energy with angle of rotation of skittle, illustrating the characterisation of stable and metastable equilibrium by a minimum of potential energy and of unstable equilibrium by a maximum. In this context gravitational energy can be equated to free energy.*

From this we see that to knock the pin over it is first necessary to increase its energy to take it over the 'energy hill' or 'energy barrier', separating the state of higher free energy from the lower *equilibrium state*. This energy increment ($\Delta G^{\ddagger}$ in Figure 2.2) may be called the *free energy of activation* for the process involved.

The above analogy illustrates some important aspects of the description of equilibrium states. According to Figure 2.1, the flat position (c) is that of lowest accessible free energy and is said to be the state of *stable equilibrium*. Position (a), although stable with respect to small disturbances, will pass over into (c) when the disturbance exceeds a critical value; it is called a state of *metastable equilibrium*. The intermediate position (b), at which the free energy is a maximum, is in principle one that could be achieved by careful balancing, but an infinitesimal disturbance in either direction will cause the skittle to fall into one or other of the energy minima; this intermediate state is one of *unstable equilibrium*.

Physico-chemical systems are of course much more complex than this, but they can nevertheless be represented in similar terms, the free energy being plotted against an appropriate 'reaction parameter'. Thus a chemical reaction is also characterised by an activation energy associated with the rate-controlling molecular mechanism involved in the reaction.

The activation energy needed for the process to occur is often provided as kinetic energy. In the case of skittles this is from the impact of the ball, which must transfer sufficient kinetic energy to tilt the pin through the critical angle θ_0 corresponding to the top of the energy barrier. In chemical systems the random impacts of colliding molecules arising from their thermal motions may provide a given molecule, or a pair of colliding molecules, with enough energy to enable reaction to occur. Alternatively the energy can come from the absorption of a photon of radiation.

The chance of the process occurring by a collision mechanism depends both on the fraction of molecules having the requisite excess energy and on the chance that they will collide. This is analogous to the calculation we could make of the probability of knocking down a skittle, which is proportional both to the chance that the throw is accurate and to the chance that the throw is powerful enough so that on impact the energy barrier is overcome. Since in physico-chemical systems the energies of molecules are distributed about a mean value according to the Maxwell–Boltzmann distribution law, there will always be a chance, however small, that the change from a metastable state to the stable state will occur, though if the barrier is very high it may take place imperceptibly slowly.

SURFACE FREE ENERGY

The discussion of the preceding section suggests that it will be useful to deal with the stability of colloids in terms of the free energy of a colloidal dispersion. It was stressed in Chapter 1 that an important characteristic of disperse systems is the large area of the interface between the particle or droplet and the surrounding medium, with the consequence that a significant proportion of the molecules are associated with the microheterogeneous regions which form the interfaces between the dispersed phase and the dispersion medium. Since the contributions which these molecules make to the thermodynamic properties of the system are different from those made by the ones within each of the bulk phases, the presence of the interface must affect the overall thermodynamic state of the system and in particular its free energy. For example, in the process of breaking a column of material of cross-sectional area A (Figure 2.3), the increase in potential energy of the system which accompanies this process is measured by the amount of work needed to separate the pieces reversibly against the forces of attraction between them (ΔW). If the process is carried out isothermally, then this is equal to the increase in free energy (see

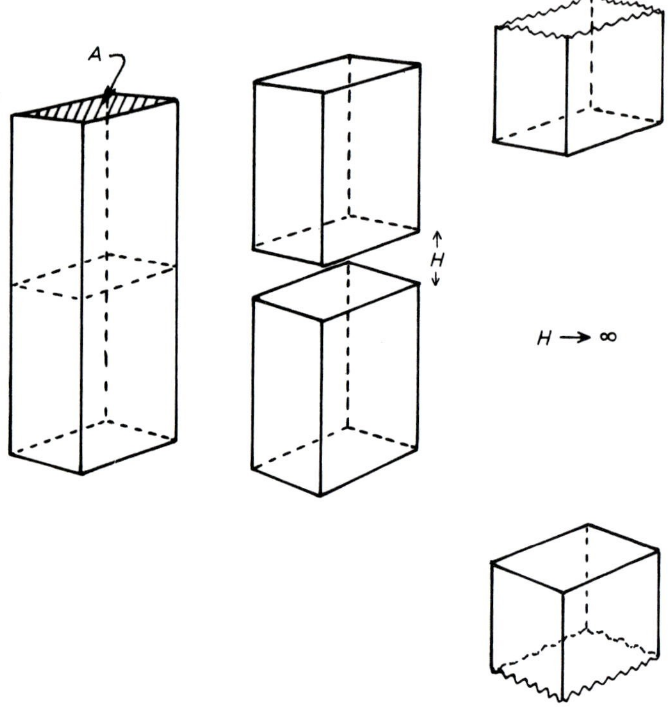

Figure 2.3 *Splitting of a column of material of cross-sectional area A to form two surfaces of total area 2A, and separation to infinity.*

the appendix at the end of this chapter). Figure 2.4(a) shows schematically the amount of work needed to separate the surfaces reversibly as a function of the distance of separation H. When the pieces are infinitely far apart the increase of free energy is proportional to the area ($2A$) of surface created and is called the *surface excess free energy*,

$$\Delta G = \Delta W = 2\sigma^0 A, \tag{2.1}$$

where the proportionality factor σ^0 is called the *surface* or *interfacial tension*.*

Initially, all the molecules in the planes between which separa-

* These terms may be used interchangeably, although surface tension usually refers to a surface between a condensed phase (solid or liquid) and a vapour or vacuum, while interfacial tension refers to that between two condensed phases.

tion is to take place are 'bulk molecules'; in the separated state they are 'surface molecules' in a quite different molecular environment and with different energies. The increase in free energy thus arises from the difference between the intermolecular forces experienced by surface molecules compared with those acting on them when they are part of the bulk material [Figure 2.5(a) and (c)].

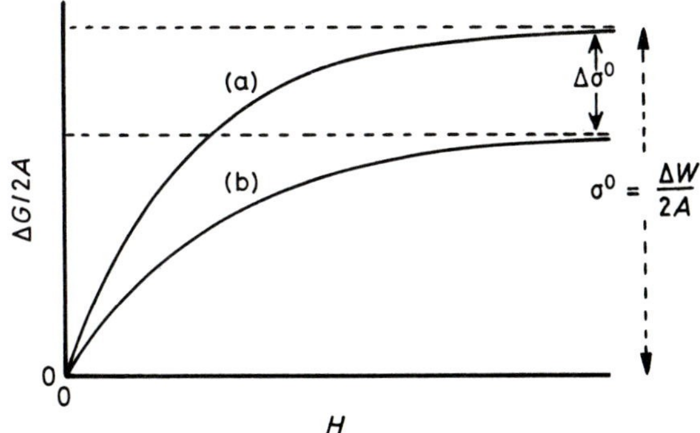

Figure 2.4 *Variation of the free energy per unit area of surface of the system shown in Figure 2.3 as a function of separation, H, of the surfaces:* (a) *in vacuo,* (b) *immersed in a fluid medium. The free-energy change per unit area at complete separation is the surface tension of the free surface σ^0, which is reduced by $\Delta\sigma^0$ on immersion. The free-energy change at intermediate separations can be regarded as the surface tension of the surfaces at that separation. Since $(d\Delta G/dH)$ is positive, the surfaces attract one another at all separations (see the appendix on page 28).*

It is seen that the surface excess free energy is not fully developed immediately since at small separations the surface molecules are still to some extent under the influence of those on the opposite faces [Figure 2.5(b)]. One may say that the surface excess free energy, and hence the interfacial tension of the surfaces, depends on their separation; we shall have occasion to use this concept in later chapters.

The above discussion refers to an idealised situation. In practice the change in surface area may be brought about by comminution of a solid or, for a liquid, by forcing it out of a nozzle to form an aerosol of liquid droplets. In both cases work has to be performed. However, it is impossible to carry out a grinding process or to form an aerosol spray reversibly, so that the increase in free energy is less than the work done in the process: part of the work is degraded by frictional processes and manifests itself as 'heat'. So these are not practicable ways of actually measuring surface free energies. However, the area of a liquid surface can be varied by forcing a drop of liquid slowly from a syringe tip, and this is the basis of one method of measuring its surface tension (see page 72). The determination of the surface free energies of solids, however, raises a number of important questions which cannot be dealt with here.

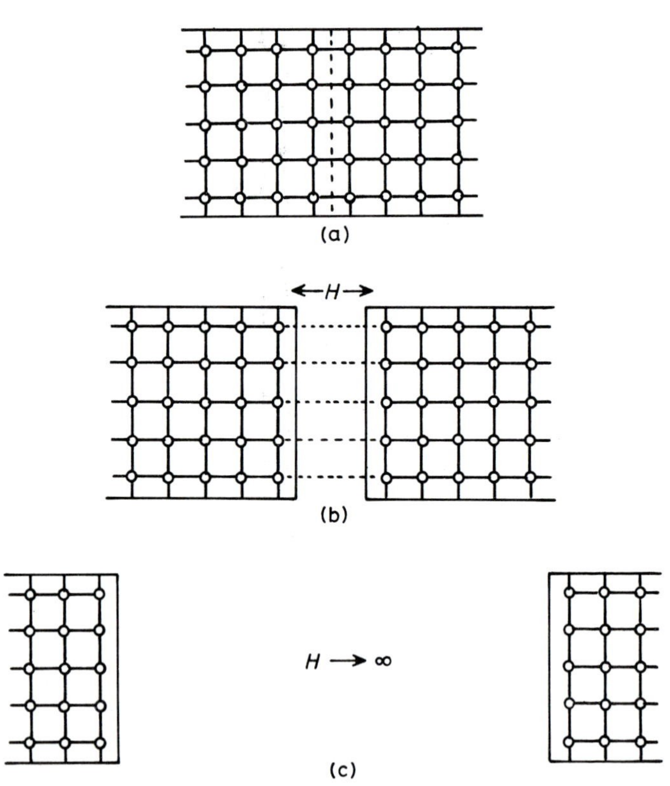

Figure 2.5 *Schematic representation of interatomic (or intermolecular) forces* (a) *in the solid before splitting,* (b) *at a separation* H *when the atoms on each surface still interact with those on the opposite surface, and* (c) *at infinite separation where the surface atoms interact only with the bulk atoms below the surface.*

The form of the curve shown in Figure 2.4(a) refers to the case in which the surfaces are formed in a vacuum or an inert gas. An attractive force acts at all separations. The free energy thus decreases as the two broken surfaces come together, and they tend to re-adhere. Indeed this tendency is manifested in practice by the 'caking' of fine powders, a problem of importance in their handling and in the design of hoppers.

The situation is somewhat different if the separation is carried out under a liquid. As we shall see later, if the space between the broken surfaces is filled with a pure liquid, the force between them is reduced and the interaction energy curve will be modified as shown in Figure 2.4(b). The free energy is reduced at all separations but the surfaces will still attract one another at all distances, albeit less strongly.

It is clear that this simple discussion of surface free energies is unable to account for colloid stability. This can only arise if there is some way of inserting an energy barrier between the states of separation and contact which prevents the metastable state from passing over into that of lower energy, *i.e.* the state of contact.

Energies have always to be measured relative to some chosen initial state. So far we have chosen zero separation as this reference level. For many purposes, however, it is more convenient to choose infinite separation as the energy zero so that Figure 2.4 takes the form shown in Figure 2.6. The curve now

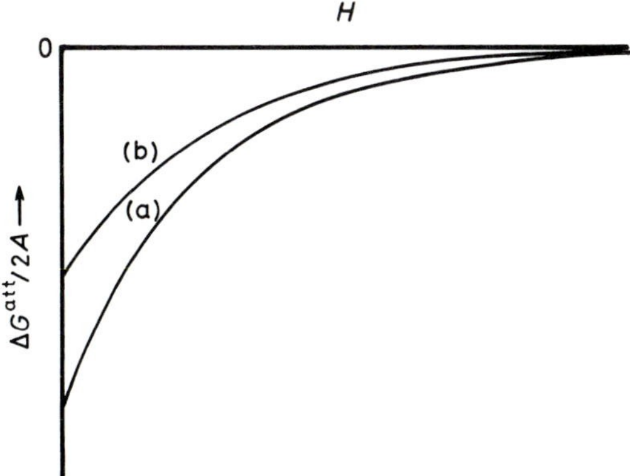

Figure 2.6 *Variation of the free energy per unit area of surface as a function of H taking as energy zero that at infinite separation, (a) and (b) as in Figure 2.4.*

represents the attractive free energy $\Delta G^{att}/2A$ and becomes increasingly negative as the surfaces approach.

REPULSIVE FORCES: THE TOTAL FREE-ENERGY CURVE

If the medium between the two surfaces is no longer a pure liquid, but is one of certain types of solution to be discussed later, then new phenomena may arise leading to a repulsive force between the surfaces. If such a repulsive force exists, then work must be done to reduce the distance between them. This is shown in Figure 2.7: the free energy arising from repulsion increases as the separation decreases.

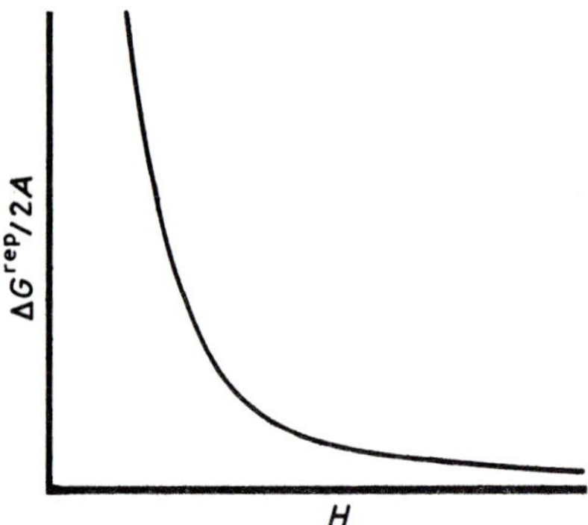

Figure 2.7 *Influence of repulsive forces on the free energy of interaction of surfaces as a function of separation, taking the energy at infinite separation as the energy zero. Work has to be done on the system to bring the surfaces together.*

It is usually assumed that the contributions to the total free energy from attractive and repulsive forces are additive, so that depending on the range and relative strengths of these two contributions a variety of total free-energy curves may result. Some typical shapes are shown in Figure 2.8. We shall leave until the next chapter a discussion of the origins of the various forces leading to surface free-energy curves such as those in Figure 2.8. Before doing so we will consider how the stability of a colloid is related to the shape of such curves. It is their contribution to the total free energy of a dispersion that is usually the most important.*

COLLOID STABILITY

From what has been said earlier, it will be clear that a colloidal dispersion represents a state of higher free energy than that corresponding to the material in bulk (Figure 2.4); passage to a state of lower free energy will therefore tend to occur spontaneously unless there is a substantial energy barrier preventing the elimination of the colloidal state. In the presence of such a barrier the system will be *metastable* and may remain in that state for a very long time.** On the other hand, if conditions are adjusted so that the energy barrier becomes negligibly small, or disappears altogether [Figure 2.8(b) and (c)], then the colloid becomes *unstable*. Thus the whole question of the preparation and stability of colloidal systems is closely tied up with the factors that give rise to free-energy barriers of adequate height to prevent the breakdown of the colloidal state. Conversely the problems of the destruction of colloids depend on our ability to change conditions so that such energy barriers are reduced or eliminated.

In the case of colloidal dispersions the energy necessary to carry a system over the energy barrier comes from the Brownian motion of the particles (see Chapter 6) which results from the random bombardment of the surface of the particles by molecules of the medium. It turns out that the average translational energy of colloidal particles undergoing Brownian motion is of the order of $(3/2)kT$ per particle, where k is Boltzmann's constant, equal to 1.38×10^{-23} J molecule^{-1} K^{-1}, and T is the absolute temperature. At 300 K two particles bring into a collision an energy of the order of 10^{-20} J. However, there is a finite probability that at a given moment a particle may have a larger or smaller energy. The chances of a collision involving a total energy of several times kT (say $10kT$) become very small so that, provided the free-energy barrier is sufficiently high, compared with kT, the dispersion will remain indefinitely in a metastable state: it is said to be

* The entropy change associated with subdivision or aggregation cannot always be ignored and is responsible for a number of special phenomena (see Chapters 9 and 11).

** Some of Faraday's original gold sols are still in existence at the Royal Institution in London.

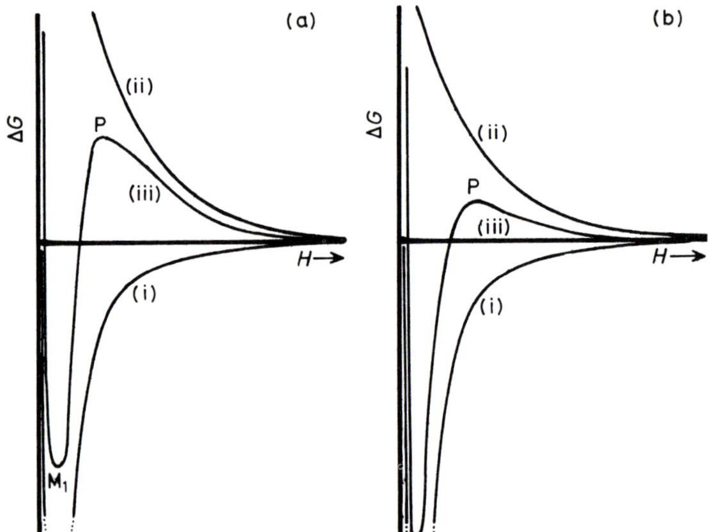

Figure 2.8 *Some possible forms of the total-interaction free energy* (iii) *resulting from a combination of attractive* (i) *and repulsive* (ii) *contributions:* (a) *states of separation and contact separated by a high energy barrier (*primary maximum, P*) arising from strong repulsive interaction,* (b) *lowering of the energy barrier by reduction of the repulsive interactions or a decrease in their range, the system being able to pass over into the* primary minimum, M_1, *if the primary maximum is*

colloidally stable.

Instability will ensue if the ratio of the barrier height to kT is reduced to say 1—2. This may arise in various ways. In principle, if the absolute height remained constant, then instability could be induced by an increase in temperature, but this is a relatively small effect. In practice the barrier height is found to be a sensitive function of several factors including the composition of the medium, temperature, and pressure. Instability and flocculation are, therefore, more often caused by the reduction in the height of the barrier resulting from a change in one or more of these factors.

A situation peculiar to colloidal systems is that in which the free-energy curve has the form of Figure 2.8(d). Here aggregation is prevented by a high energy barrier, but preceding this is a relatively shallow minimum – called a *secondary minimum* to distinguish it from the deep *primary minimum*. If the depth of the

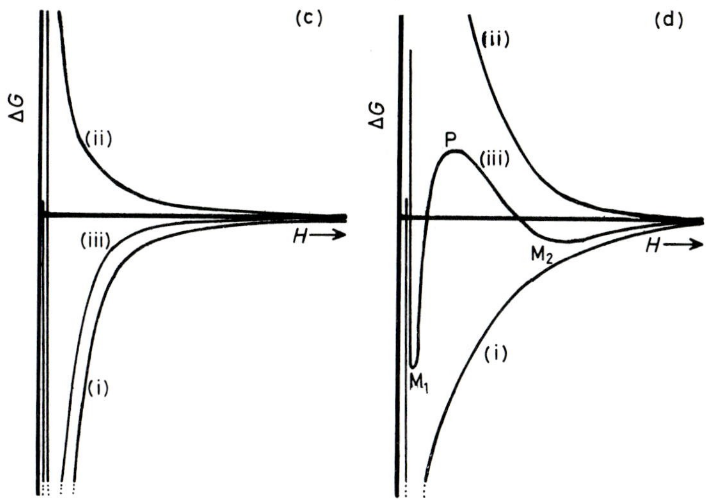

lowered to a few kT, (c) *elimination of the energy barrier, the system being able to pass over into the* primary minimum, M_1, *without any activation energy,* (d) *it is sometimes found that a shallow minimum exists (*secondary minimum, M_2*) before the primary maximum is reached. In these diagrams account is also taken of the short-range repulsive forces when the surfaces are almost in contact (Born repulsion). Figures 2.4 and 2.6 refer to 'hard' interactions where on contact the repulsive force becomes infinite.*

secondary minimum is of the order of a few kT, then small relatively weakly bound aggregates (*flocs*) can form. These have a limited lifetime, and a kinetic equilibrium may be set up between single particles and flocs, which is often described as *weak flocculation* or *secondary minimum flocculation*. An important feature of such a situation is that, although the flocs are sufficiently stable not to be completely dissociated by Brownian motion, they may disintegrate under externally applied hydrodynamic forces, such as vigorous stirring. Some practical implications of this phenomenon will be discussed in Chapter 14.

In the following chapters we shall first seek to understand the origin of free-energy curves of the various types illustrated in Figure 2.8 and then apply these ideas to a discussion of the formation and stability of some typical systems.

So far we have limited our attention to one particular kind of colloid, namely particulate dispersions. In these the driving force for aggregation or dispersion is the gradient of the free-energy curve. However, the surface free energy of a system is a function both of the surface tension and of the interfacial area:

$$G = \sigma A. \tag{2.2}$$

The processes we have considered so far may therefore be regarded as taking place at constant surface area:

$$\mathrm{d}G = A\mathrm{d}\sigma. \tag{2.3}$$

However, there are many examples in which the driving force is the change in the surface area of interfaces, the surface tensions remaining essentially constant:

$$\mathrm{d}G = \sigma\mathrm{d}A. \tag{2.4}$$

As we shall see, this is the case in the destruction of emulsions by the coalescence of emulsion droplets. In more complex systems, which we shall not discuss, colloid stability is controlled by changes in both surface tension and area:

$$\mathrm{d}G = \mathrm{d}(\sigma A) = \sigma\mathrm{d}A + A\mathrm{d}\sigma. \tag{2.5}$$

APPENDIX

We consider the sequence illustrated in Figure 2.3. At a given separation H the external force which has to be applied to hold the two parts at this distance is equal and opposite to the force of attraction between the parts $\mathcal{F}^{\mathrm{att}}$. By convention, $\mathcal{F}^{\mathrm{att}}$ is given a negative sign so that the work done reversibly on the system when the separation is increased by $\mathrm{d}H$ is $-\mathcal{F}^{\mathrm{att}}\mathrm{d}H$, which is positive, in accordance with the usual thermodynamic convention. The work done on the system in the whole separation process, carried out reversibly, is then given by

$$\Delta W = -\int_0^\infty \mathcal{F}^{\mathrm{att}}\mathrm{d}H.$$

If the process is carred out isothermally, then ΔW measures the increase in free energy, ΔF, between the initial and final states, while the slope of the curve of ΔF as a function of H, $[\partial(\Delta F)/\partial H]_T$, is equal to $-\mathcal{F}^{\mathrm{att}}$.

That ΔW is not equal to the total energy change, ΔU, is readily understood since, in general, the process will be accompanied by a flow of heat, ΔQ, into or out of the system to maintain constant temperature. By the first law of thermodynamics

$$\Delta U = \Delta W + \Delta Q,$$

while, according to the second law, in a reversible change

$$\Delta Q = T\Delta S,$$

where ΔS is the entropy change accompanying the separation process. It follows that

$$\Delta W = \Delta U - T\Delta S,$$

which by definition is equal to ΔF.

Note that in the present context no distinction need be made between the energy U and the enthalpy H, nor between the Helmholtz free energy F and the Gibbs free energy G. In the following chapters we shall use the Gibbs free energy.

Chapter 3

Why are Colloidal Dispersions Stable? II Interparticle Forces

INTRODUCTION

In Chapters 1 and 2 we saw that *in vacuo* or in an inert atmosphere the free energy of two bodies increases as the separation between their surfaces increases so that they will attract one another and the natural tendency will be for them to come together spontaneously; if the two bodies are colloidal particles they will tend to aggregate. We have, however, indicated that under certain circumstances the mutual free-energy curves of colloidal particles may exhibit a maximum (Figure 2.8) so that when the particles approach the free energy increases initially, *i.e.* the particles repel one another, until at the maximum in the curve this repulsion changes over to an attraction. The system thus has to surmount a free-energy barrier, which if sufficiently high (compared with kT) will keep the particles apart so that the dispersion will be colloidally stable. In this chapter we shall discuss the various factors which contribute to the shape of interparticle free-energy curves and hence control the stability of dispersions.

INTERMOLECULAR FORCES

Our ultimate aim is to understand the interactions between colloidal particles. However, since these are attributed to the summation of the interactions between individual molecules in the two particles, we must begin by discussing the origin and nature of intermolecular forces.

The existence of attractive forces between non-polar molecules has been recognised since the classical work of van der Waals (1873), but their origin was not understood until London (1930) showed how they could be calculated by a quantum mechanical discussion of the interaction between fluctuating dipoles, arising from the motions of the outer electrons on the two molecules.*

According to the theoretical equations, these attractive forces increase in magnitude as the molecules approach one another, and at a separation of r between the nuclei of atoms (or the centre of mass of roughly spherical molecules) they are proportional to the inverse seventh power of the separation:**

$$\mathcal{F}^{\text{att}} = -A/r^7. \tag{3.1}$$

The work done in separating reversibly a pair of atoms or molecules from a distance d to infinity is given by

$$\Delta W = -\int_d^\infty \mathcal{F}^{\text{att}}\mathrm{d}r = A\int_d^\infty (1/r^7)\mathrm{d}r = A/6d^6 = A'/d^6. \tag{3.2}$$

If we choose the energy at infinite separation as the energy zero, then the free energy of attraction between a pair of atoms or molecules at a separation d is

$$\Delta G^{\text{att}} = -\Delta W = -A'/d^6 \tag{3.3}$$

and is shown schematically as curve (a) in Figure 3.1. The constant A' is related to the nature of the individual molecules through certain quantum mechanical properties, which can be expressed approximately in terms of experimental quantities.

One expression, used by London, gives A' for the interaction of two identical atoms or molecules in the form

$$A' = (3/4)h\nu\alpha^2, \tag{3.4}$$

where α is the polarisability of the atom or molecule, h is Planck's constant (6.63×10^{-34} J s), and ν is a characteristic frequency identified with that corresponding to the first ionisation

* The term 'dispersion forces', often used to describe these forces, will not be used here in order to avoid confusion with the term 'dispersion' used to describe colloidal systems. We shall call them London–van der Waals forces. The van der Waals forces between polar molecules include other contributions which have to be added to the London forces.

** We adopt the convention of giving a negative sign to attractive forces and a positive sign to repulsive forces. Note that the external force needed to balance the attractive force is $-\mathcal{F}^{\text{att}}$ (see page 28). *N.B.* A here is a constant, not area.

potential and consequently lying in the ultraviolet. For two unlike molecules, 1 and 2,

$$A'_{12} = (3/2)h\left(\frac{\nu_1\nu_2}{\nu_1 + \nu_2}\right)\alpha_1\alpha_2. \tag{3.5}$$

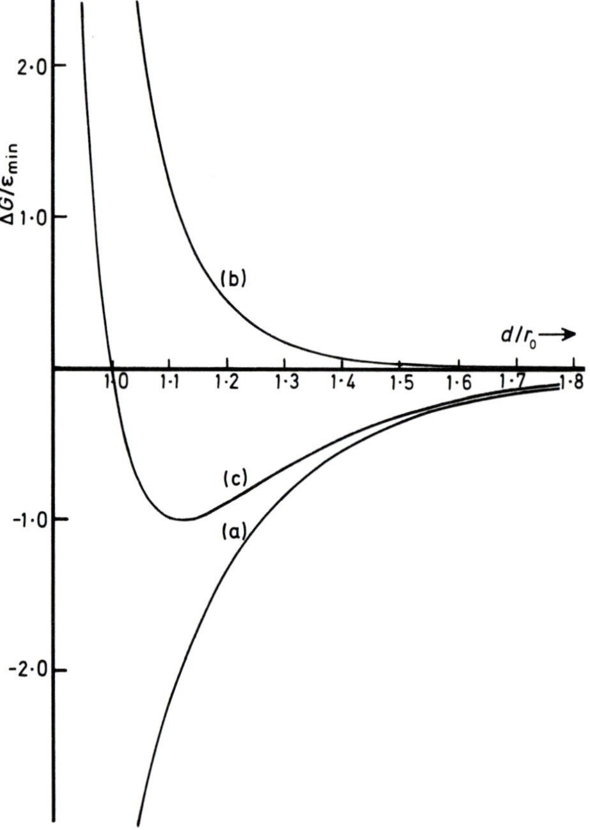

Figure 3.1 *Potential energy of interaction, $\Delta G(d)$, between two molecules a distance d apart: (a) van der Waals attraction, (b) Born repulsion, (c) resultant intermolecular potential.*

The attractive force increases and the free energy becomes increasingly negative as the atoms approach one another. However, at close distances their electron clouds begin to interact. If the electrons are in non-bonding orbitals, this gives rise to a repulsive force and an increase in free energy which becomes effectively infinite when the electron clouds interpenetrate. This repulsive force (sometimes called the *Born repulsion*) is expected theoretically to have the form

$$\mathcal{F}^{\text{rep}} = B\mathrm{e}^{-ad}, \tag{3.6}$$

where a and B are constants, leading to a contribution to the intermolecular potential at a distance d apart of

$$\Delta G^{\text{rep}} = (B/a)\mathrm{e}^{-ad}. \tag{3.7}$$

The algebraic handling of problems involving intermolecular potentials is greatly simplified if the repulsive potential is represented by the approximate expression*

$$\Delta G^{\text{rep}} = B'/d^{12}, \tag{3.8}$$

which is shown as curve (b) in Figure 3.1. The total potential energy of interaction between a pair of atoms or molecules is assumed to be the sum of the contributions from attractive and repulsive forces so that, as shown in curve (c) of Figure 3.1,

$$\Delta G = \Delta G^{\text{rep}} + \Delta G^{\text{att}} = (B'/d^{12}) - (A'/d^{6}). \tag{3.9}$$

This is usually known as the Lennard–Jones potential.** Instead of expressing the equation in terms of the parameters A' and B' it is often more convenient to use the depth of the potential well ε_{min} and the distance at which the potential is zero, r_0, when

$$\Delta G = 4\varepsilon_{\text{min}}[(r_0/d)^{12} - (r_0/d)^{6}]. \tag{3.10}$$

INTERPARTICLE FORCES

One method of calculating the potential energy of interaction between two particles is to assume that every molecule in one particle interacts with each molecule in the other according to a Lennard–Jones potential and that the total free energy of interaction is obtained by summing the contributions from all possible pairs of molecules [Figure 3.2(a)]. The repulsive contribution can be neglected except for those molecules on opposing surfaces. The

* Now that these problems are more frequently solved by computer, the advantages of equation (3.8) over (3.7) have largely disappeared.

** A similar expression in which the exponents 12 and 6 were left unspecified as *n,m* had been used empirically much earlier by Mie.

simplest case to handle mathematically is that of two hard, flat, effectively infinite plates a distance H apart [Figure 3.2(b)], for which it was shown by Hamaker that, *for unit area* of surface,

$$\Delta G^{\text{att}} = -A_{\text{H}}/(12\pi H^2), \tag{3.11}$$

where A_{H} is called the Hamaker constant. Its value is closely related to A' of equation (3.4) and is given by

$$A_{\text{H}} = (3/4)h\nu\alpha^2\pi^2q^2 = A'\pi^2q^2, \tag{3.12}$$

where q is the number of molecules in unit volume of particles. A similar calculation for two equal-sized spheres [Figure 3.2(c)] of

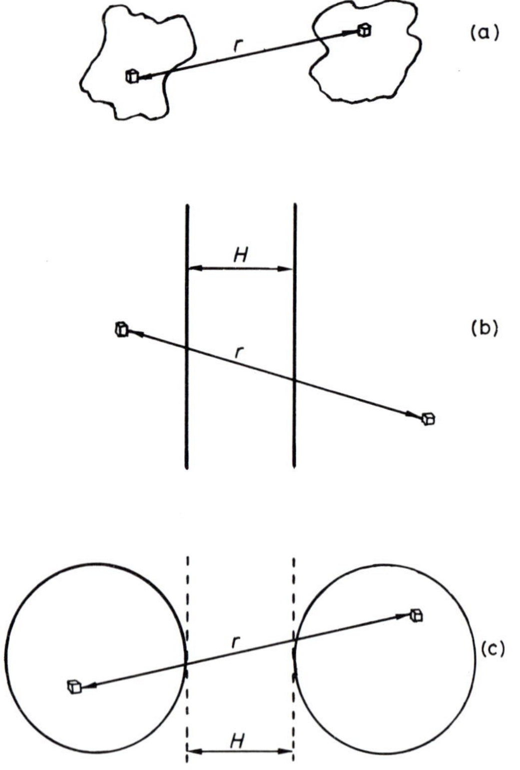

Figure 3.2 *Interaction between elements of volume dV containing qdV molecules, which on summation for all pairs of volume elements gives the total-interaction free energy between the two bodies:* (a) *particles of arbitrary shape,* (b) *two parallel semi-infinite plates a distance H apart,* (c) *two spheres whose surfaces are a distance H apart.*

radius a gives, as an approximation when the spheres are close ($H/a \ll 1$):

$$\Delta G^{\text{att}} = -(A_{\text{H}}a/12H)[1 + (3/4)H/a + \text{higher terms}]. \quad (3.13)$$

For many purposes it is sufficient to neglect all but the first term in brackets.

Both formulae (3.11) and (3.13) emphasise a very important point, namely that the energy of attraction between *particles* falls off very much more slowly than that between *single molecules*: the long range of interaction between colloid particles plays a crucial role in determining their properties.

When close-range repulsion is taken into account, the total free-energy curves for particles interacting through London–van der Waals forces have the forms already shown schematically in Figure 2.8.

Because the forces between colloidal particles have a much longer range than those between individual molecules, an extra factor has to be considered in discussing interparticle forces. In the London theory the interaction between fluctuating dipoles is supposed to occur instantaneously. However, the electromagnetic field set up by a moving electron is propagated at the velocity of light. There is thus a delay between the wave being sent out by one molecule, its arrival at a neighbouring molecule, and the response of that molecule being received by the first. By this time, if the molecules are far apart, the electronic state of the first molecule may have changed. For molecules the range of the interactions, and hence the time-scale, is so short that this effect can be ignored. For particles, however, it becomes important. Typically for separations greater than 10 nm this so-called *retardation effect* is significant and causes the interaction to fall off more rapidly than it does at closer separations. Initially, as we have seen, the energy of interaction between two plates falls off in proportion to H^{-2}; at greater distances, when the forces are 'retarded', the relation changes to a proportionality to H^{-3}. While the retardation effect is important in quantitative theories of colloid interaction, it does not influence the general principles which are the subject of this book.

EFFECT OF THE INTERVENING MEDIUM

The expression given above for the Hamaker constant for the

interaction between two particles applies only to the case in which the particles are in a vacuum; it is also adequate in the presence of an inert gas and so can be used, for example, in discussing the interaction between aerosol particles.

Particles immersed in a liquid medium experience a smaller attractive force. The calculation of such forces and their relationship to the nature of the particles and of the dispersion medium has been, and continues to be, a major theoretical problem. Important advances towards its solution have been made in recent years, and they will be mentioned briefly in Chapter 15. For the present purposes it is sufficient to use a relatively simple approximate equation according to which the appropriate Hamaker constant to be employed when two particles of material 1 are separated by a medium 2 is given by

$$A_{\mathrm{H}} = [A_{10}^{1/2} - A_{20}^{1/2}]^2, \tag{3.14}$$

where A_{10} is the Hamaker constant for interaction between two particles of material 1 through a vacuum, and similarly for A_{20}. The important qualitative deduction from this, which is confirmed by more sophisticated theories, is that the more closely similar the disperse phase and the dispersion medium the closer will be the values of $A_{10}^{1/2}$ and $A_{20}^{1/2}$ and hence the smaller the value of A_{H} for particles of 1 immersed in 2. Because equation (3.14) involves the square of the difference between A_{10} and A_{20}, the same will be true for particles of 2 immersed in 1. The form of the interaction curve between particles immersed in a liquid thus has the same form as in Figure 2.4(b) but with a different value of the Hamaker constant. So far, therefore, theory predicts that colloidal particles will attract one another at all distances of separation down to that of the primary minimum; we still have to seek the origin of the repulsive forces that give rise to colloid stability.

ELECTROSTATIC FORCES: THE ELECTRICAL DOUBLE LAYER

In Chapter 1 it was stated that the particles in most colloidal dispersions in aqueous media carry an electric charge. We also learned that the stability of such dispersions is very sensitive to the addition of electrolytes. Evidence for the existence of charges on particles comes from the phenomenon of electrophoresis, which will be dealt with in Chapter 6. Meanwhile we accept their presence and consider their origin and their role in stabilising colloids.

Surfaces may become electrically charged by a variety of mechanisms, the more important of which are the following:

(i) *Ionisation of surface groups*. If the surface contains acidic groups, their dissociation gives rise to a negatively charged surface [Figure 3.3(a)]; conversely, a basic surface takes on a positive charge [Figure 3.3(b)]. In both cases the magnitude of the surface charge depends on the acidic or basic strengths of the surface groups and on the pH of the solution. The surface charge can be reduced to zero (at the *point of zero charge, p.z.c.*) by suppressing the surface ionisation by decreasing pH in case (a) or by increasing the pH in case (b). Many metal oxides exhibit amphoteric behaviour in that both positively and negatively charged surfaces can be obtained by varying the pH.

(ii) *Differential solution of ions from the surface of a sparingly soluble crystal*. For example, when a silver iodide crystal is placed in water, solution occurs until the product of ionic concentrations equals the solubility product: $[Ag^+][I^-] = K_s = 10^{-16}$ $(mol\,dm^{-3})^2$. If equal amounts of Ag^+ and I^- ions were to dissolve, then $[Ag^+] = [I^-] = 10^{-8}\,mol\,dm^{-3}$ and the surface would be uncharged. In fact silver ions dissolve preferentially, leaving a negatively charged surface [Figure 3.3(c)]. If Ag^+ ions are now added (in the form say of silver nitrate solution), the preferential solution of silver ions is suppressed and the charge falls to zero when $[Ag^+] = 10^{-5.5}$ (or $pAg = -\log[Ag^+] = 5.5$); further addition leads to a positively charged surface, since it is now iodide ions that are preferentially dissolved [Figure 3.3(d)]. If I^- ions had been added instead (say as potassium iodide solution), then the negative charge would have been increased since the preferential solution of the Ag^+ ions would have been increased. The surface charge thus depends on the relative concentrations of Ag^+ and I^- ions, the p.z.c. occurring at $pAg = 5.5$. (An alternative, but equivalent, description of the above sequences can be given in terms of the relative adsorptions of Ag^+ and I^- ions at the crystal surface.)

(iii) *Isomorphous substitution*. A clay may exchange an adsorbed, intercalated, or structural ion with one of lower valency, producing a negatively charged surface. For example, Al may replace Si in the surface, tetrahedral, layer of a clay, producing a negative surface charge [Figure 3.3(e)]. In this case the p.z.c. can be reached by reducing the pH, the added H^+ ions combining with the negative charges on the surface to form OH groups.

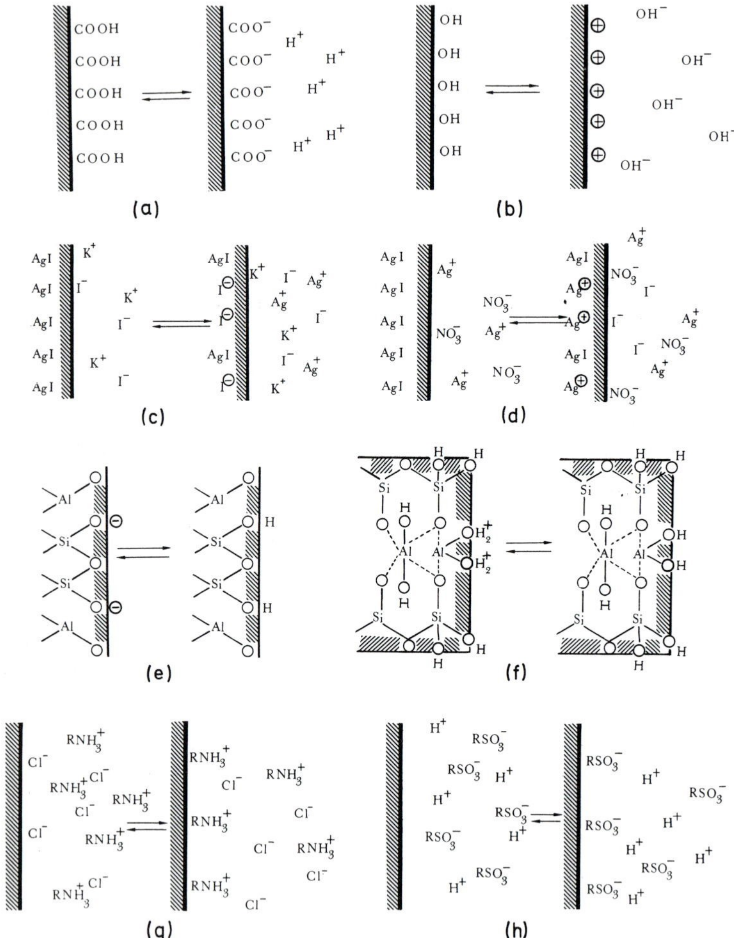

Figure 3.3 *Origin of surface charges by:* (a) *ionisation of acid groups to give a negatively charged surface,* (b) *ionisation of basic groups to give a positively charged surface,* (c) *differential solution of silver ions from a AgI surface,* (d) *differential solution of iodide ions from a AgI surface,* (e) *isomorphous substitution in a clay surface to give a negatively charged surface,* (f) *breaking a clay crystal to give a positively charged edge,* (g) *specific adsorption of a cationic surfactant,* (h) *specific adsorption of an anionic surfactant.*

(iv) *Charged crystal surfaces*. It may happen that, when a crystal is broken, surfaces with different properties are exposed. Thus in some clays (*e.g.* kaolinite), when a platelet is broken, the exposed edges contain AlOH groups which take up H^+ ions to give a positively charged edge [Figure 3.3(f)]. This may coexist with negatively charged basal surfaces, leading to special properties (see Chapter 13). In this case there will be no single p.z.c., but each type of surface will have its own value.

(v) *Specific ion adsorption*. Surfactant ions may be specifically adsorbed, leading, in the case of cationic surfactants, to a positively charged surface [Figure 3.3(g)] and, in the case of anionic surfactants, to a negatively charged surface [Figure 3.3(h)].

To examine the way in which these charges affect the properties of colloids, it is first necessary to say a little about fundamental electrostatic theory.

The fundamental law of electrostatics – Coulomb's law – expresses the inverse square law of force between two electric charges q_1 and q_2 separated by a distance d in a vacuum in the form:

$$\mathcal{F} = q_1 q_2/(4\pi\varepsilon_0 d^2), \tag{3.15}$$

where ε_0 is the permittivity of free space (vacuum). If q_1 and q_2 have the same sign, the force is a repulsion; if they are of opposite sign, the force is an attraction.*

In the presence of a material medium surrounding both charges, the force is reduced by a factor $\varepsilon_r = \varepsilon/\varepsilon_0$, the *relative permittivity* (or *dielectric constant*) of the medium. The work done in bringing two charges together from infinite separation to a distance d in a medium of permittivity ε is therefore given by

$$\Delta G^{el} = \Delta W = -\int_{\infty}^{d} \mathcal{F} \mathrm{d}h = q_1 q_2/(4\pi\varepsilon d) \tag{3.16}$$

and measures the electrical free energy of the system relative to that at infinite separation.

The charge q_1 may be thought of as producing an *electric field* at a point d, such that the work needed to bring a *unit charge* to this point is $q_1/(4\pi\varepsilon d)$; this is called the *electrical potential* at d due to

* This is the form taken in the S.I. system of units. The original form (in e.s.u. units) may be more familiar to some readers: $\mathcal{F} = q_1 q_2/d^2$. Note the sign convention on page 28.

the charge q_1 and is denoted by ψ. The work done to bring up a charge q_2 to this point is $q_2\psi$ and is positive if q_2 is positive and conversely.

To enable us to discuss the electrostatics of electrolyte solutions we need to introduce another fundamental principle – *Boltzmann's distribution law* – which relates the probability of particles being at a given point at which they have a potential energy, or free energy, ΔG, relative to some chosen reference state. This probability may be expressed in terms of the average concentration, c, at the point considered relative to that, c^0, at the reference level, taken as the zero of energy. If the temperature is T, then

$$c = c^0 \exp(-\Delta G/kT). \tag{3.17}$$

The simplest example of this is, of course, the distribution of gas molecules in the Earth's atmosphere. The potential energy of a gas molecule of mass m (relative to its value at the surface of the Earth) is $mg\Delta h$, where Δh is the height of the point considered above the surface. Thus

$$\ln(c/c^0) = -mg\Delta h/kT, \tag{3.18}$$

which indicates that (if the temperature of the atmosphere were constant and independent of height) the concentration of gas molecules and hence the atmospheric pressure should decrease logarithmically with height, which is broadly true. The same equation applies to the distribution of colloidal particles in the Earth's gravitational field, as shown by Perrin in his classical work on gamboge particles (see Chapter 6).

Applied to the electrostatic case, Boltzmann's law indicated that, if at some point in an electrolyte solution there is an electrical potential ψ, then in the region of that point the concentration of positive ions will be

$$c(+) = c^0 \exp(-z_+ e\psi/kT), \tag{3.19}$$

where z_+ is the valency of the positive ion, e the elementary (protonic) charge, and c^0 the concentration of positive ions in a region where $\psi = 0$. For negative ions similarly

$$c(-) = c^0 \exp(+z_- e\psi/kT). \tag{3.20}$$

In this region there will therefore be an imbalance of electrical charges, which in the case where $z_+ = z_- = 1$ is

$$c(+) - c(-) = c^0[\exp(-e\psi/kT) - \exp(+e\psi/kT)]. \tag{3.21}$$

If the region under consideration is close to a negative ion, then ψ will be negative and $[c(+) - c(-)]$ will be positive. This means that around a negative ion there will be an excess of positive charge; this is called a *charge cloud* or *ionic atmosphere*. Since the electrolyte solution as a whole is electrically neutral, the integral $[c(+) - c(-)]\mathrm{d}v$ taken over the whole volume (v) of the solution outside the negative charge must exactly balance that charge. This picture is the starting point for the Debye–Hückel theory of electrolytes in which the contribution of electric charges to the free energy of an electrolyte solution is calculated.

In colloid science we are particularly interested in the ionic atmosphere which is developed around a charged colloid particle, rather than around a single ion. In this context it is usual to call this ionic atmosphere an *electrical double layer*. The charge on the particle is distributed over its surface and is just balanced by the total charge in the double layer in which there is an excess of oppositely charged ions (*counter-ions*) (Figure 3.4).

The concept of an electrical double layer was introduced by

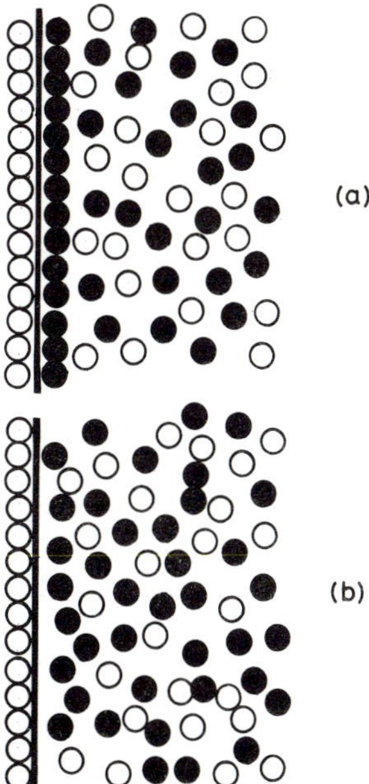

Figure 3.4 *The electrical double layer:* (a) *according to the Helmholtz model,* (b) *the diffuse double layer resulting from thermal motion.* ○ *positive charge,* ● *negative charge.*

Helmholtz, who envisaged an arrangement of charges in two parallel planes as shown in Figure 3.4(a), forming in effect a 'molecular condenser'. However, thermal motion causes the counter-ions to be spread out in space, forming a *diffuse double layer* in which the local concentration is determined by equation (3.21), Figure 3.4(b).

The case most fully studied is that of a plane surface carrying a uniformly distributed charge. The electrical potential in the solution then falls off exponentially with distance from the surface (Figure 3.5):

$$\psi = \psi^0 \exp(-\kappa z). \quad (3.22)$$

Thus at a distance $1/\kappa$ the potential has dropped by a factor of $(1/e)$. This distance may be used as a measure of the extension of the double layer and is often loosely called the *thickness of the double layer*. According to the theoretical equations it has the value $1/\kappa = [\varepsilon kT/e^2\Sigma c_i z_i^2]^{1/2}$* and is identical with the parameter introduced in the Debye–Hückel theory of electrolytes in which $1/\kappa$ is identified with the radius of the ionic atmosphere. Of particular importance in colloid science is the fact that the thickness of the

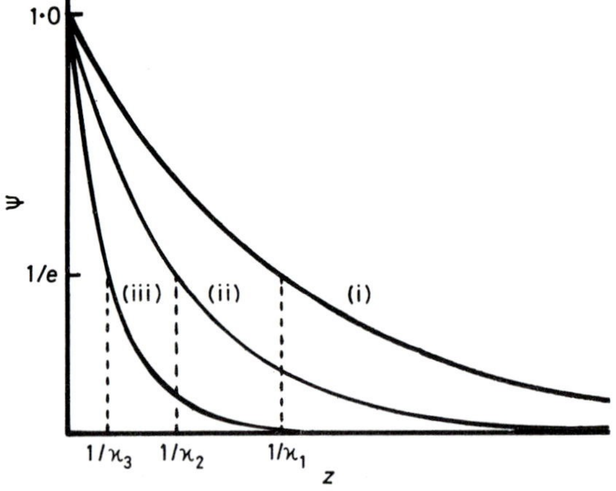

Figure 3.5 *Variation of electrical potential with distance from the surface and its dependence on electrolyte concentration:* (i) *low,* (ii) *medium,* (iii) *high electrolyte concentration. $1/\kappa$ is called the* thickness of the double layer.

* *N.B.* Here c_i is the number of ions m^{-3}; if c_i is expressed in $mol\,m^{-3}$, then $1/\kappa = [\varepsilon RT/F^2\Sigma c_i z_i^2]^{1/2}$, where F is the Faraday constant.

double layer depends markedly on the ionic concentration as shown in Figure 3.5: as the concentration increases, the thickness of the double layer decreases rapidly. Thus in an aqueous solution of a 1:1 electrolyte at 25 °C the values of $1/\kappa$ are: at

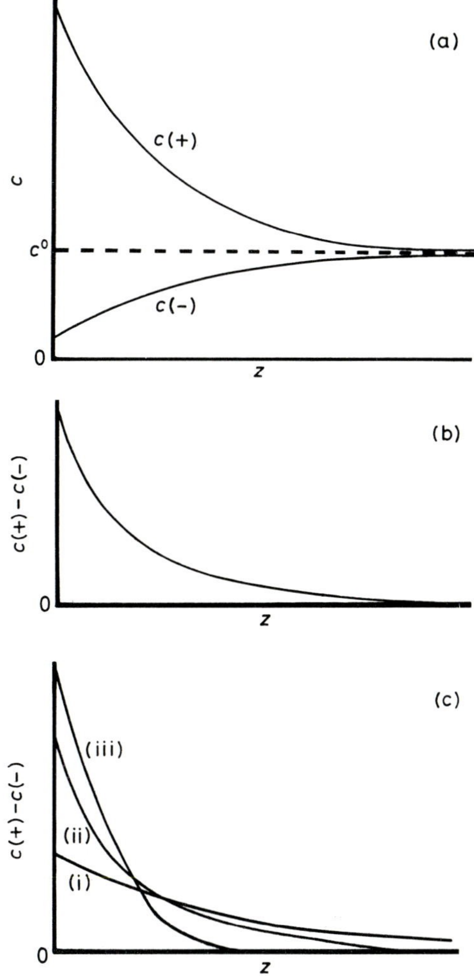

Figure 3.6 *Charge distribution in the diffuse double layer. Variation with distance from the surface of:* (a) *the concentrations of positive and negative ions in the neighbourhood of a negatively charged surface,* (b) *the net excess charge,* (c) *dependence of the net charge distribution on the ionic strength of the solution, for the case in which the surface charge is constant. Curves* (i), (ii), *and* (iii) *refer to increasing ionic strength.*

10^{-4} mol dm^{-3} 30.4 nm, at 10^{-3} mol dm^{-3} 9.6 nm, at 10^{-2} mol dm^{-3} 3.0 nm, and at 10^{-1} mol dm^{-3} 0.96 nm.

It follows from equation (3.22) that the local ionic concentrations vary with the distance from the surface in the way depicted in Figure 3.6(a), while the local charge density $[c(+) - c(-)]$ is shown in Figure 3.6(b). The area under this curve is equal to the charge on the surface. From these considerations we arrive at the picture of a charged colloid particle as being enclosed in a sheath in which charges of opposite sign to that on the particle predominate. The distance over which this imbalance of local charges persists is dependent on the electrolyte concentration, as shown in Figure 3.6(c) for the case in which the surface charge is constant, *i.e.* the areas under the various curves are constant.

In colloid science we have to consider not only the nature of a single double layer (a subject of importance in surface electrochemistry) but also the way in which the double layers surrounding two colloid particles interact with one another when they come together.

When two similarly charged colloid particles, with their associated double layers, move toward one another they will begin to 'feel' one another's presence as soon as any appreciable overlap of the 'tails' of the charge distributions occurs [Figure 3.7(a)]. This interaction will develop steadily as they approach. The resulting repulsion between the particles may be pictured, and interpreted in detail, in various ways and at various levels of sophistication (see Chapter 9). One simple view is that the two ionic charge clouds, of the same sign, repel one another in much the same way as the electron clouds around atoms and molecules. Alternatively, the double layers may be thought of as screening the charges on the colloid particles from one another. Seen from a sufficient distance, a charged particle with its neutralising double layer 'looks' like an uncharged entity. At closer separations the screening is incomplete and each particle 'sees' the other as a partially charged particle of the same sign, as a consequence of which the particles repel one another. The full solution of the electrostatic problem still cannot be achieved without mathematical approximation and physical simplifications, but the general picture is clear. The general form of the double-layer repulsive energy and its dependence on electrolyte concentration is that shown in Figure 3.7(b).

We have now identified one mechanism by which colloid

particles may repel one another. It has the important properties that it is dependent on the charge on the particle (and hence on the electrical potential at its surface) and on the electrolyte concentration, both of which affect the range of the repulsive force. We shall come back later in the chapter to consider the consequences of adding this repulsive potential to that arising from van der Waals forces. Before doing so we deal briefly with a second phenomenon which can lead to repulsion between particles.

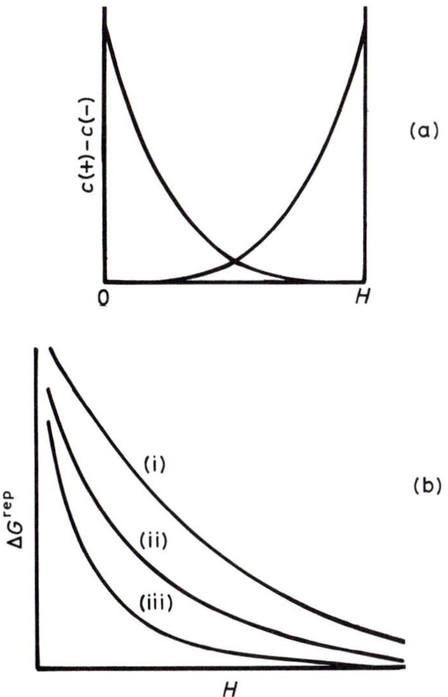

Figure 3.7 (a) *Overlapping of the two double layers when two charged particles approach.* (b) *Double-layer repulsive energy as a function of particle separation. Curves* (i), (ii), *and* (iii) *refer to increasing electrolyte concentration.*

'STERIC REPULSION': THE EFFECT OF ADSORBED OR ANCHORED LAYERS

It has been known for a very long time, empirically, that the addition of certain natural biopolymers such as gelatine can stabilise colloidal dispersions. The earliest example is probably the use of natural gums by the Chinese and Egyptians 2000—3000 years ago to prepare inks. Their action was attributed by Faraday and later authors to the formation of a protective sheath around each colloid particle which prevented flocculation. The phenomenon of stabilisation by this means was called *protection*.

In recent years systematic studies using a wide variety of synthetic polymers have led to a considerable elucidation of the detailed mechanism of stabilisation of this kind, which has been given the name *steric stabilisation*.

Adsorbed layers can affect interparticle forces in two ways. First they can influence the van der Waals attractive forces, and secondly they can give rise to a repulsion between the particles. We consider two cases.

(i) *Anchored (or effectively anchored) layers*. If an adsorbed layer is compact and is not displaced on collision between particles, then the presence of this layer has the effect of preventing the centres of the particles from coming closer than $2(a + \delta)$, where a is the radius of the particles and δ the thickness of the adsorbed layer (Figure 3.8). The particles may be regarded as essentially hard spheres with an effective (collision) diameter of $2(a + \delta)$. In this simple case the attractive van der Waals potential can be calculated by an extension of the pair-wise summation procedure described earlier and illustrated in Figure 3.8. Two extreme situations are possible. If the surface layer has properties closely similar to those of the dispersion medium, then for a given centre-to-centre distance the attraction will be unaffected by the presence of the surface layer. However, since the centre-to-centre distance cannot fall below $2(a + \delta)$, the attractive potential at contact will be less than that for an uncoated particle ($\delta = 0$) [Figure 3.9(a)]. On the other hand, if the coating is closely similar to the core material, the coated particle will behave simply as a larger particle, and in accordance with equation (3.13) it will have a greater attractive potential than the uncoated particle [Figure 3.9(b)].

It follows that it is possible to reduce the depth of the potential well on contact by choosing for the adsorbed layer a polymer which, although strongly attached to the surface, has properties matching those of the dispersion medium. If a sufficient reduction is achieved, *e.g.* by having δ large enough and so reducing the well depth to a few kT, then Brownian motion (see Chapter 6) may keep the system in a dispersed state.

This picture is, however, incomplete. The adsorbed layer of polymer is not usually dense enough to enable it to behave as a hard surface. The polymer chains will extend out into the medium to an extent which depends on how they interact with the medium [Figure 3.10(a)]: the more closely similar chemically are the polymer segments and the medium the more open the surface structure. On collision between particles the polymer

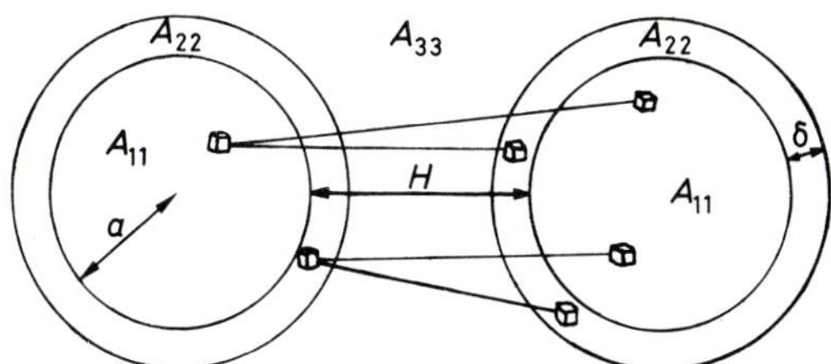

Figure 3.8 *Interaction between two particles carrying an anchored polymer layer of thickness δ. The interaction energy is calculated by summing the contributions between the pairs of elements as indicated.*

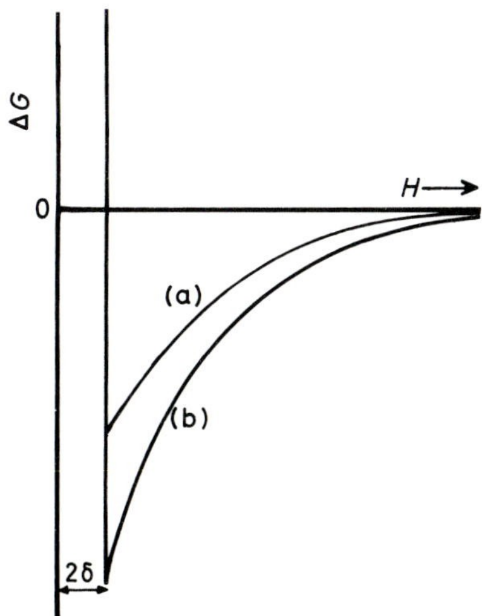

Figure 3.9 *Interaction potential curves corresponding to Figure 3.8.* (a) *If the polymer has properties closely similar to the dispersion medium,* $A_{22} \simeq A_{33}$; (b) *if the polymer has properties closely similar to the particle,* $A_{11} \simeq A_{22}$.

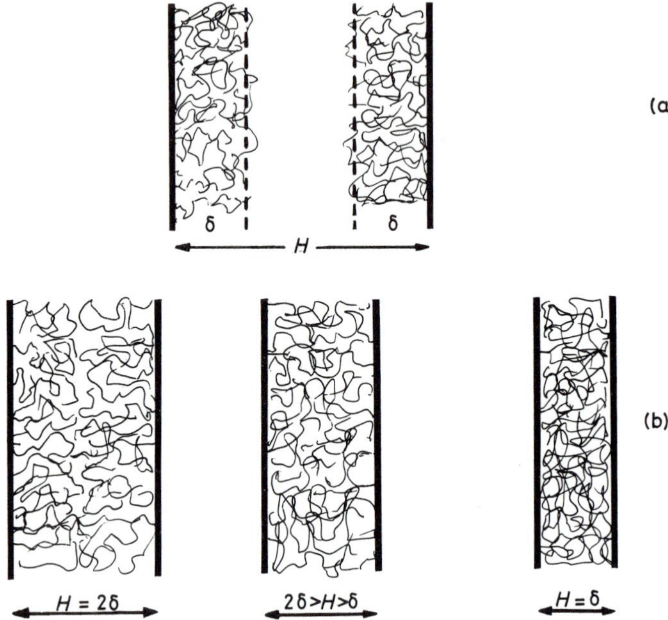

Figure 3.10 *Origin of the steric repulsion between particles:* (a) *particles far apart, $H \gg 2\delta$,* (b) *successive stages of overlap of the polymer layers as H goes from 2δ to δ. $\delta \simeq \alpha(rl^2)^{1/2}$, where r is the number of segments in the polymer, l is the length of each segment, and α is an expansion coefficient ($\alpha = 1$ for a random coil).*

chains will interpenetrate [Figure 3.10(b)]. This has two consequences. First, the local density of polymer segments increases. Osmotic effects (see Chapter 6) will then cause the medium to diffuse into the region between the surfaces to reduce the segment concentration and so to drive the surfaces apart. Secondly, because the segments are linked together in a polymer chain, this increased concentration will constrain the chains, leading to a reduction in the number of configurations they can adopt. This implies a reduction, ΔS, in the entropy of the system and hence an increase, $|T\Delta S|$, in the free energy. This contribution to the intermolecular potential is called the *entropic repulsion* term. In this case the centre-to-centre separation of the particles can fall below $2(a + \delta)$, but as it does so the potential rises steeply. This is shown schematically in Figure 3.11(a). The magnitude of the repulsion arising from the presence of the adsorbed layer clearly depends on the density with which it covers the surface: the more thinly it is spread the smaller its effectiveness in preventing the particles from approaching one another. It follows that, as the density decreases, the total-interaction potential curves can take on the series of forms shown in Figure 3.11(b), and eventually the adsorbed layer is unable to prevent the system from passing over into the primary minimum. If ΔS is constant, the repulsion free energy arising from this effect increases with increasing temperature, so that systems stabilised in this way may be expected to become more stable as the temperature is raised. However, other factors, to be discussed in greater detail in Chapter 9, show that, although steric stabilisation is initially favoured by an increase in temperature, this tendency may be reversed at higher temperatures.

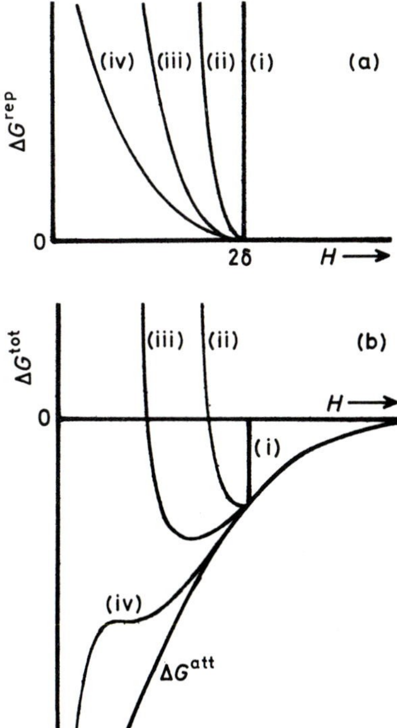

Figure 3.11 (a) *Contribution to the repulsive free energy for sterically stabilised systems:* (i) *hard surface,* (ii)—(iv) *decreasing density of the adsorbed layer, at constant* δ. (b) *Total-interaction free energy for systems in* (a). *If the adsorbed layer becomes too sparse* (iv), *it may not be able to prevent coagulation into the primary medium.*

To exploit this phenomenon it is necessary to devise methods of preparation of colloidal dispersions which lead to particles having anchored to them a layer of polymer of a nature different from that of the particle, as closely similar as possible to the medium, and of sufficient chain length to give an adequately large value of δ. The polymer may be attached by specific chemical bonds, by chain entanglements, or by strong adsorption. In the latter case it is advantageous to employ a block copolymer of the form A_nB_m or $B_mA_nB_m$ in which the A-chains are strongly adsorbed by the particle while the B-chains are 'soluble' in the medium.

(ii) *Reversibly adsorbed substances*. In the above discussion we have supposed that no desorption of the adsorbed layer occurs during the time involved in a collision. This is often an inadequate assumption. If the substance is adsorbed reversibly, and equilibrium is maintained during a collision, then somewhat different considerations apply. It is a fundamental result of surface thermodynamics that adsorption of a substance lowers the surface tension at that surface. Thus aqueous soap solutions in which soap molecules are adsorbed at the liquid/gas interface have a lower surface tension than water. Let us return to the splitting process illustrated in Figure 2.3, but now with the surfaces immersed in a solution. Suppose that, once the surfaces have been separated, one component of the solution is preferentially adsorbed, then the free-energy curves will be modified as shown in Figure 2.4. At infinite separation the surface tension of the surfaces will be reduced by an amount $\Delta\sigma = (\sigma^0 - \sigma)$, the magnitude of which depends on the amount adsorbed.*

In general, the adsorption will depend on the separation of the surfaces: its contribution to σ will vary with H. It turns out, as will be explained in Chapter 5, that, if the adsorption increases as the surfaces come together, adsorption effects enhance the attraction, while, if Γ decreases as H decreases, adsorption effects oppose the attraction and may lead to an overall repulsion.

* According to the Gibbs adsorption isotherm (see page 67),

$$d\sigma = -RT\Gamma d\ln a,$$

where Γ is the adsorption on unit area of surface and a is the activity of the adsorbed component in the solution. Thus on going from pure solvent ($x = a = 0$) to solution of activity a:

$$\sigma^0 - \sigma = -RT\int_{-\infty}^{\ln a}\Gamma d\ln a.$$

OTHER FACTORS

The presence of polymers, either adsorbed on colloidal particles or free in solution, can lead to other interesting effects. For example, if a high-molecular-weight polymer is present at low concentration, remote segments of a polymer chain may be adsorbed on separate particles, causing them to be drawn together (*bridging flocculation*). The presence of an excess of non-adsorbed polymer can also result in flocculation (*depletion flocculation*). These and other special cases will be discussed in Chapter 9.

THE TOTAL-INTERACTION POTENTIAL CURVE

We have now considered briefly the main contributions to the total free energy of interaction between two colloid particles. The most general equation for the total free-energy difference (ΔG) between particles at infinite separation and at a separation H is obtained by adding these contributions:

$$\begin{aligned}\Delta G = {} & \Delta G^{\text{att}} \text{ (van der Waals)} + \Delta G^{\text{rep}} \text{ (short range)} \\ & + \Delta G^{\text{rep}} \text{ (electrostatic)} + \Delta G^{\text{rep}} \text{ (steric)} \\ & + \Delta G \text{ (other effects)}. \end{aligned} \tag{3.23}$$

In practice it is not necessary to consider all these contributions simultaneously, except in certain special cases. We shall here deal with the two simpler situations in which the long-range repulsive potential arises *either* from electrostatic *or* from steric contributions. Figures 3.12(a) and 3.12(b) show schematically the total free-energy (or potential) curves in these two cases. Particular attention is drawn to the way in which the double-layer repulsion depends on the ionic strength of the medium: the curves may show a high repulsive barrier at low ionic strengths, a so-called secondary minimum at intermediate ionic strengths, and a negligibly small barrier, or none at all, at higher ionic strengths. In the same way the form of the steric repulsion is determined by the nature of the interactions between the adsorbed polymer chains and the solvent. A repulsive barrier of variable range and a minimum of variable depth can result, depending on the solvent and the temperature.

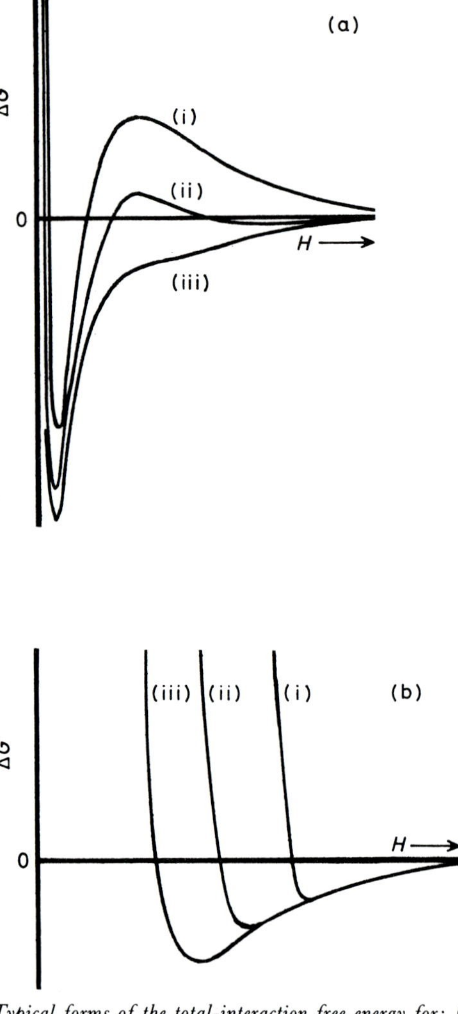

Figure 3.12 *Typical forms of the total-interaction free energy for:* (a) *electrostatically stabilised systems [curves* (i), (ii), *and* (iii) *refer to increasing electrolyte concentration]*, (b) *sterically stabilised systems [curves* (i), (ii), *and* (iii) *refer to constant density of polymer chains, but decreasing* δ*, arising from decreasing values of* α*]*.

SUMMARY

In the last two chapters we have explained in general terms some of the main theoretical ideas upon which much of colloid science is based. In later chapters these concepts will be developed in greater detail and applied to a discussion of the properties of a variety of colloidal systems.

It is important to stress again that what we have done so far is to draw together several aspects of physical chemistry and show how they are relevant to the study of colloids.

Chapter 4

How are Colloidal Dispersions Prepared?

INTRODUCTION

It is evident from the description of colloidal systems in Chapter 1 that there are two fundamentally different ways in which colloidal dispersions can be formed, either by breaking down bulk matter to colloidal dimensions (*dispersion methods*) or by building up molecular aggregates to colloidal sizes (*condensation* or *nucleation methods*). We learned in Chapters 2 and 3, however, that such dispersions will only remain in a metastable or colloidally stable state if conditions are arranged so that the repulsive forces are large enough to prevent the particles from coagulating or flocculating.

In this chapter we shall consider these two routes to colloidal dispersions in turn. The formation of gels and association colloids will be dealt with separately in later chapters.

DISPERSION METHODS

Comminution

We have already discussed in Chapter 2 the energy changes associated with the breaking of a column of material into two parts which are then separated to infinity. If this process is continued until the solid is broken down into particles of colloidal size, the increase in free energy accompanying this process will be given by the product of the surface area produced and the surface tension of the surfaces so formed. It follows that the ease with which a powder can be produced by grinding depends on the surface tension of the solid. If the comminution process could be continued until the material were reduced to single molecules, then the work done would correspond to the energy of sublimation or evaporation of the bulk material. It is not surprising therefore that both empirical and theoretical relationships have been proposed between the surface tensions and the energies of sublimation or evaporation of solids and liquids.

If the grinding process takes place in an inert atmosphere, there will be a strong tendency for the particles to adhere to one another – the caking of dry powders is a common problem in practice.

Since immersion of a solid surface in a liquid results in a lowering of surface tension (see Figure 2.4), grinding is often most conveniently carried out under a liquid. Furthermore, by controlling the composition of the liquid by the addition of, for example, *dispersing agents* a stable dispersion may be achieved. The nature and mode of action of dispersing agents or stabilisers have been hinted at in Chapter 3 and will be discussed in more detail later. Meanwhile, we note that in industrial applications dispersion methods are in many instances the preferred procedures and that a wide range of dispersing agents is commercially available.

Emulsification

The formation of emulsions by breaking down one liquid in the presence of another may also be achieved by mechanical means. In some instances simple shaking or stirring may be sufficient; in others it is necessary to apply very strong hydrodynamic forces as is done in commercial 'colloid mills' or 'emulsifiers'. In these a coarse mixture of the two liquids is forced under pressure through a narrow gap and sometimes, by having relative motion between the surfaces forming the gap, is subjected simultaneously to shear.

The success of an emulsifying process depends in large measure on the interfacial tension between the two liquids. This can be modified by the addition of appropriate *emulsifying agents*, whose mode of action we shall again discuss later. In favourable cases, where the interfacial tension is very low, the energy needed to form an emulsion is correspondingly small and may be provided by the thermal motion of the molecules. *Spontaneous emulsification* is an important phenomenon, made use of, for example, in the dispersion of agricultural chemicals in water and, potentially, in oil recovery. In many instances (*e.g. microemulsions*) the dispersions so produced are thermodynamically stable.

As mentioned earlier, it is not uncommon for emulsion droplets to have diameters well above the accepted colloid size range. However, even in this case colloid principles are of fundamental importance when one considers the problem of droplet coalescence in the destruction of emulsions (*de-emulsification*). Here the medium between two approaching droplets is progressively thinned, and before coalescence can occur it is reduced to a film of colloidal thickness (see Chapter 12).

Suspension and Aerosol Methods

An important technique is that in which it is the precursor of the final colloidal particle that is reduced to a colloidal size. Thus a liquid reactant may be emulsified and then caused to react to form a colloidal dispersion of solid particles whose particle size distribution is related to that of the emulsion precursor. The commonest application of this method is in *suspension polymerisation*,* in which an emulsion of monomer droplets, stabilised by a surfactant, is polymerised by adding an initiator which is soluble in the monomer. Polymerisation occurs within the monomer droplet, leading to the formation of a *polymer latex*.

The use of aerosol particles for the same purpose is a very recent development. For example, a titanium(IV) ethoxide aerosol reacted with water vapour yields amorphous spherical titanium dioxide particles, while exposure of droplets of *p*-t-butylstyrene to a vapour initiator yields particles of the corresponding polymer.

CONDENSATION METHODS: NUCLEATION AND PARTICLE GROWTH

In principle one can prepare a colloidal particle from a molecular species by forming molecular complexes of increasing size until the colloid size range is reached, and this is the way in which very many dispersions are formed. Usually the molecular (or atomic) species are formed by a chemical reaction, and being virtually insoluble in the dispersion medium they aggregate into particles of increasing size. Among familiar examples are the formation of colloidal sulphur by the interaction of sodium

* Not to be confused with *emulsion polymerisation* (see page 60).

thiosulphate solution with acid, of colloidal gold by the reduction of gold chloride, and of colloidal silver halides by the reaction of an alkali halide with silver nitrate solution (see Appendix I).

This process is, however, not as simple as this brief description implies, and stable colloidal dispersions are obtained only if the conditions are properly controlled.

In the first place, passage from the molecular state to a particle involves passing through the intermediate size range of 1—10 nm in the process of *nucleation*. We can see rather generally what this involves in the following way; a more rigorous alternative treatment will be mentioned in Chapter 11.

As a rough approximation we can think of a cluster, or *embryo*, of n molecules as being made up of a core of n^{b} molecules having the properties of the bulk solid, surrounded by n^{s} surface molecules (Figure 4.1). The free energy of the cluster, $g(n)$, can thus be expressed in terms of bulk (g^{b}) and surface (g^{s}) molecular free energies:

$$\begin{aligned} g(n) &= n^{\mathrm{b}}g^{\mathrm{b}} + n^{\mathrm{s}}g^{\mathrm{s}} \\ &= (n^{\mathrm{b}} + n^{\mathrm{s}})g^{\mathrm{b}} + n^{\mathrm{s}}(g^{\mathrm{s}} - g^{\mathrm{b}}). \end{aligned} \tag{4.1}$$

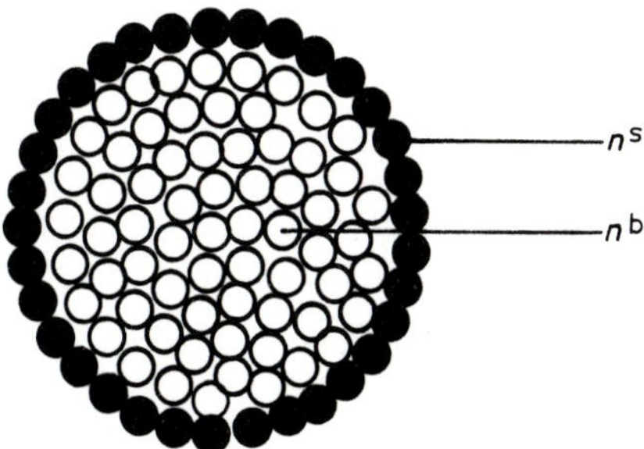

Figure 4.1 *An embryo pictured as a core of n^b molecules having the same properties as bulk material and a surface layer of n^s molecules.*

We have already seen (Chapter 2) that the surface tension of a pure substance can be equated to the difference per unit area in free energy between bulk and surface molecules, so that in this approximate equation (4.1) we may write*

* For the present purposes we can neglect the difference between the Helmholtz free energy (F) and the Gibbs free energy (G). In this form of the theory the translational entropy of the embryo is also neglected.

$$g(n) = ng^{\mathrm{b}} + \sigma A, \tag{4.2}$$

where

$$\sigma = (g^{\mathrm{s}} - g^{\mathrm{b}})n^{\mathrm{s}}/A \tag{4.3}$$

and A is the surface area. Now for a sphere $A \propto n^{2/3}$ and the molecular free energy of the bulk solid is equal to its molecular chemical potential μ^{b} (*i.e.* the conventional molar chemical potential divided by Avogadro's constant). The free energy of the cluster is thus of the form

$$g(n) = n\mu^{\mathrm{b}} + \sigma b n^{2/3}, \tag{4.4}$$

where b is a geometric factor.

The formation of a cluster from individual molecules of a substance A present in bulk solution at a mole fraction x may be represented as

$$n\mathrm{A} \rightarrow \mathrm{A}_n, \tag{4.5}$$

for which the free-energy change ΔG, per mole of A_n formed, is given by

$$\Delta G = g(n) - n\mu, \tag{4.6}$$

where μ is the molecular chemical potential of monomers. If the solution is dilute,

$$\mu = \mu^{\ominus} + kT\ln x, \tag{4.7}$$

where $\mu^{\ominus}$ is the standard chemical potential in solution. Thus, inserting (4.4) and (4.7) into (4.6),

$$\Delta G = [n\mu^{\mathrm{b}} + \sigma b n^{2/3}] - n[\mu^{\ominus} + kT\ln x], \tag{4.8}$$

which rearranges to

$$\Delta G = n[\mu^{\mathrm{b}} - \mu^{\ominus} - kT\ln x] + \sigma b n^{2/3}. \tag{4.9}$$

Now for a saturated solution $x = x^{\mathrm{sat}}$ and $\mu^{\mathrm{b}} = \mu^{\ominus} + kT\ln x^{\mathrm{sat}}$, so that equation (4.9) becomes

$$\Delta G = -nkT\ln[x/x^{\mathrm{sat}}] + \sigma b n^{2/3}, \tag{4.10}$$

while

$$(\partial\Delta G/\partial n) = -kT\ln[x/x^{\mathrm{sat}}] + \tfrac{2}{3}\sigma b n^{-1/3}. \tag{4.11}$$

The ratio x/x^{sat} is called the *supersaturation* of the solution. Figure 4.2 shows ΔG as a function of n. At n_{c} the maximum in the curve, $\partial(\Delta G)/\partial n$, is zero. Thus for $n < n_c$ the addition of molecules to the embryo is disfavoured and the embryo will tend to disintegrate, while for values of $n > n_c$ growth is accompanied by a decrease in free energy and is thus a spontaneous process. The embryo containing just n_c molecules is called the *critical nucleus*. Whether or not critical nuclei are formed in the course of thermal fluctuations depends on the height of the free-energy barrier: if it is high compared with kT (the thermal energy), the chance of forming such nuclei is negligible and no particles will be formed.

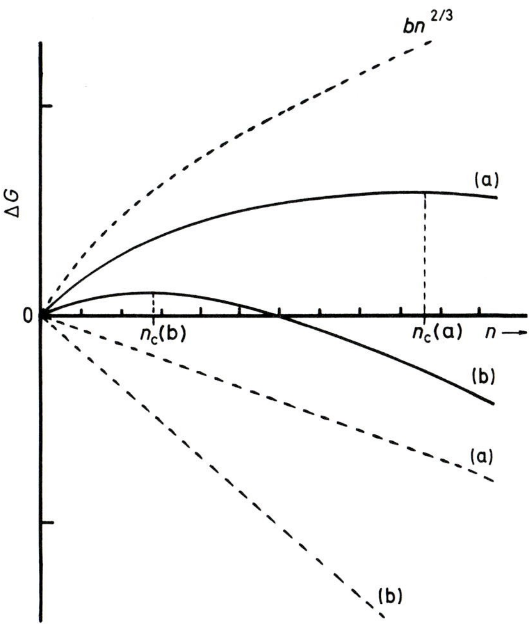

Figure 4.2 *The free energy of an embryo as a function of its size. The two terms in equation (4.7) are shown dotted. Curves* (a) *and* (b) *refer respectively to a lower and higher supersaturation; both n_c and the height of the barrier decrease as x/x^{sat} increases.*

However, the form of the curves in Figure 4.2 depends on the degree of supersaturation of the solution: as the supersaturation increases, the height of the barrier and the size of the critical nucleus both decrease, as shown in Figure 4.2(a) and (b). Eventually a concentration is reached at which the barrier is low enough for the rate of production of critical nuclei by spontaneous fluctuations to increase dramatically, and these nuclei begin to grow. This concentration is called the *critical supersaturation*.

In many cases, of course, nucleation signals the onset of the growth of embryos to macroscopic size. Familiar examples are the condensation of a supersaturated vapour and the induction of crystallisation of a solid by the addition of seed crystals (larger than the critical nucleus) or by scratching the side of the vessel to encourage nucleation. The condition for the formation of a colloidal dispersion is that the supply of molecules must run out while the particles are still in the colloidal size range. This is favoured by (i) having dilute solutions and (ii) ensuring that a very large number of nuclei are produced in as short a time as possible.

In the chemical preparations mentioned above the solubilities of the products are so low that, even when dilute solutions are used, the critical supersaturations are far exceeded.

In addition to ensuring that conditions are such that nucleation and growth of particles to colloidal size are controlled, it is also essential that an appropriate stabilisation mechanism is operative. In the case of colloids formed in aqueous media from electrolytes this stabilisation usually arises from double-layer interaction, while in non-aqueous media it is necessary to employ some form of steric stabilisation.

EMULSION AND DISPERSION POLYMERISATION

Processes of major industrial importance which depend on nucleation and growth processes include those of emulsion and dispersion polymerisation. These differ from suspension polymerisation discussed earlier. Once again the monomer, which is slightly soluble, is dispersed as emulsion droplets, but now in contrast to suspension polymerisation the initiator is insoluble in the monomer but soluble in the continuous phase. Polymerisation therefore occurs in the continuous phase, the droplets of monomer serving as a source of monomer. When the growing polymer in solution reaches a critical molecular weight, nucleation occurs and insoluble polymer particles are formed. These may then grow by the addition of polymer molecules or by direct reaction between surface free radicals and monomer in solution. Depending on the particular system involved, an emulsifying agent may be needed to stabilise the monomer emulsion, while the resulting polymer latex may be stabilised either by a surface charge or by adsorption of a steric stabiliser.

PREPARATION OF MONODISPERSE COLLOIDS

It is becoming of increasing interest from both theoretical and practical standpoints to be able to prepare monodisperse colloids of controlled size. From the above discussion it follows that to do this it is necessary for a large number of critical nuclei to be formed in a short interval of time, nucleation then to cease, and those nuclei already formed to grow simultaneously at the same rate. This can be achieved, at least in principle, if the reaction producing the precipitating species takes the concentration rapidly to somewhat above the critical supersaturation. The ensuing spontaneous nucleation then reduces the concentration to below the critical supersaturation, so that no further nuclei are formed. Growth of the existing nuclei, rather than the production of fresh nuclei, is now the preferred process, and provided the rate of supply of the precipitating species by chemical reaction is not greater than the rate of growth of the particles the original nuclei will continue to grow. This sequence of events is illustrated in Figure 4.3.

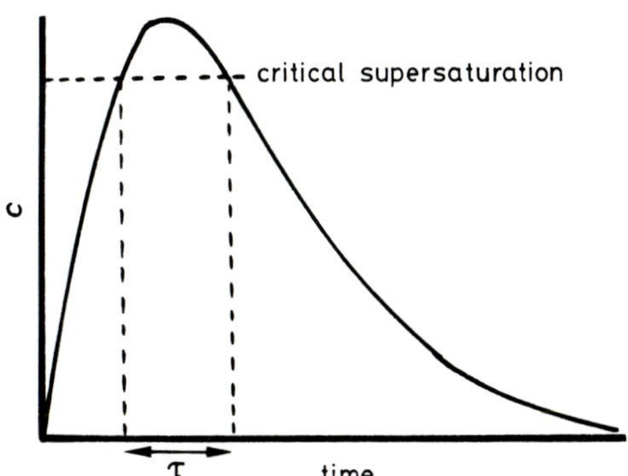

Figure 4.3 *Variation in the concentration of the precipitating species during the formation of a monodisperse colloid. The species is produced by chemical reaction until its concentration exceeds the critical supersaturation. It is then lowered by the formation of nuclei. Depletion of the solution by precipitation then reduces the concentration to below the critical supersaturation and no further nuclei are formed. The existing nuclei grow uniformly. The shorter the time τ the more uniform the final dispersion.*

It may be shown that under these conditions the spread of particle sizes (arising from the spread of time – τ in Figure 4.3 – during which nuclei are formed) becomes sharper with time. It is indeed well known that the monodispersity of colloids formed under comparable conditions improves as the particle size increases.

Methods based on the above general principles have been developed for the preparation of monodisperse colloidal dispersions of a variety of metal oxides and hydroxides and of polymer latices. The latter have played a major role in the understanding of colloids and their properties.

Chapter 5

What is the Role of Surface Chemistry? Surface Tension and Adsorption

INTRODUCTION

In the preceding chapters we have seen how one can in a general way understand colloid formation and stability in terms of the variation of free energy with the separation between the surfaces of two particles. When this free energy is measured with respect to the state in which the two surfaces are in contact, it may be identified with the surface or interfacial tension (see Figure 2.4). A major contribution to this surface tension arises from the van der Waals attractive forces between the particles. It turns out, however, that the surface tension, and hence the force between the two surfaces, is also strongly influenced by the adsorption of molecules at the surfaces.

A full discussion of this problem requires a detailed analysis in thermodynamic terms. Here we present only a general account, but the interested reader will find fuller proofs of the important equations in Appendix III.

ADSORPTION

When two phases are in contact there is a transition region of molecular dimensions in which the composition changes from that of one phase to that of the other. In the case of a solid in contact with a vacuum or inert gas, the transition region is unlikely to be thicker than one or two molecular layers and is characterised by local surface irregularities shown schematically in Figure 5.1. For a pure liquid in contact with its vapour a more gradual and smooth transition occurs, extending up to five or six molecular diameters (Figure 5.2). For a liquid mixture/vapour interface a local concentration profile exists for each component of the solution (Figure 5.3). In the case of fluid phases the thickness of the interfacial region increases with increase in temperature until it disappears at the critical temperature of the system. A similar diagram (Figure 5.4) can be drawn for a solid/liquid interface.

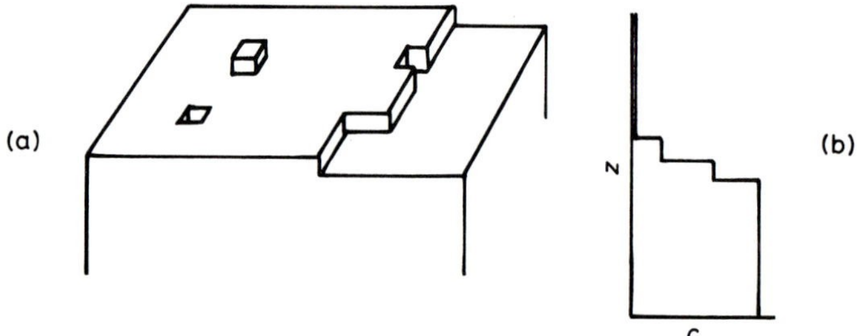

Figure 5.1 (a) *Surface of a solid showing surface imperfections.* (b) *Stepped concentration profile corresponding to* (a).

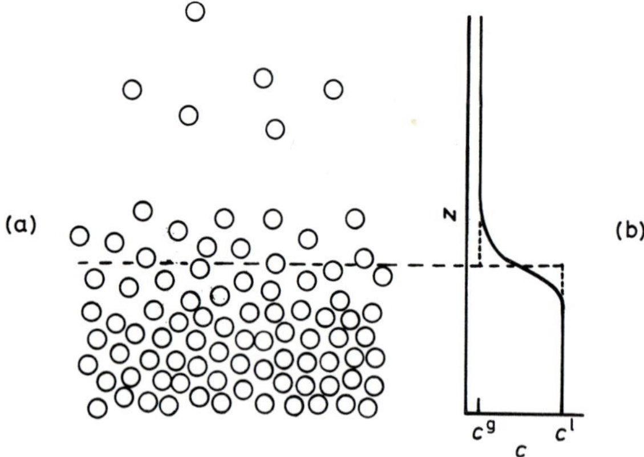

Figure 5.2 (a) *Surface of a liquid showing change of concentration from* c^l *to* c^g *on passing through the interface.* (b) *Concentration profile at a liquid/vapour interface.*

An immediate consequence of this situation is that the total amount of a particular component in a system of given volume and interfacial area will be determined by the shape of the local concentration profile. However, experimental methods of determining the shape of this profile are at an early stage of development and so far are limited to polymer solutions. Currently all that one can do is to investigate the overall macroscopic, stoichiometric, consequences of the presence of the profile at the interface. Various ways of quantifying this have been suggested, and a choice between them is largely a matter of taste since they lead in the end to essentially identical results. The course adopted by Willard Gibbs is the one we shall employ. An attempt will be made to dispel some of the supposed difficulties associated with its use.

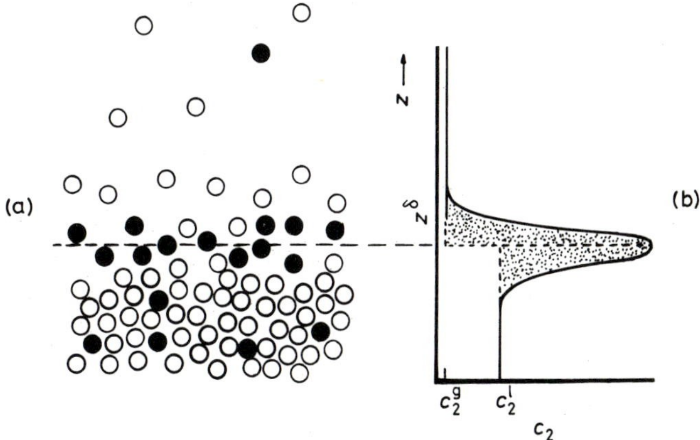

Figure 5.3 (a) *Surface of a solution of which one component is preferentially adsorbed at the liquid/vapour interface.* (b) *Concentration profile of component 2 on passing through the interface.*

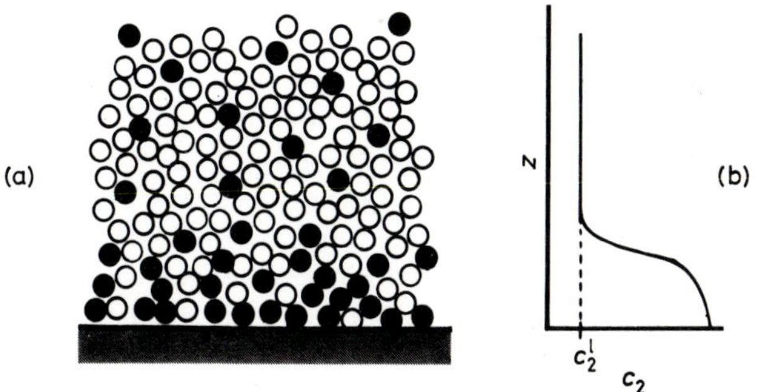

Figure 5.4 (a) *Preferential adsorption of one component of a solution at a solid/solution interface.* (b) *Concentration profile of component 2 at a solid/solution interface.*

The basic idea is very simple. Let us try to do some 'book-keeping'. We first consider the amount of a given substance, i, which we would expect to find in a volume V^α of phase α in which the concentration, uniform throughout this volume, is c_i^α, together with a volume V^β of phase β in which the concentration is c_i^β. This would be simply $(c_i^\alpha V^\alpha + c_i^\beta V^\beta)$. However, since the local values of c_i vary on passing through the interface, there will be, in general, a different amount, n_i, of i actually present. The difference

$$n_i^\sigma = n_i - (c_i^\alpha V^\alpha + c_i^\beta V^\beta) \tag{5.1}$$

is called the *surface excess amount of i* which is present because of the interface; it may also be called the amount of *i adsorbed* at the interface. The problem lies in deciding exactly how to identify the separate volumes V^α and V^β, *i.e.* in deciding on the location of the surface (called a *Gibbs dividing surface*) defining the interface between them: all we know is that the sum $V^\alpha + V^\beta$ is equal to the total volume V. The value to be attributed to n_i^σ is critically dependent on the location chosen for this surface. It turns out that one way of choosing the dividing surface which leads to equations having a simple form is to follow Gibbs in locating it at a position that reduces to zero the adsorption of one chosen component (usually the solvent denoted by 1). This is illustrated in Figure 5.5. This leads to the definition (see Appendix III) of the *relative adsorption of i with respect to component 1*, which when related to unit area of interface is denoted by $\Gamma_i^{(1)}$.

It is important to realise that the Gibbs 'model' *does not* imply that the surface excess is concentrated at the Gibbs dividing surface: this is clearly physically impossible since molecules have a finite size and cannot occupy a mathematical surface. What the Gibbs method does is to recognise the existence of concentration profiles such as those shown in Figure 5.3, whose exact form cannot yet be measured experimentally, and to provide a method of expressing the observable consequences of their existence.

Equations relating the relative adsorption to observable quantities are given in Appendix III. However, the direct measurement of surface excesses using these equations is not always easy. Thus it is extremely difficult to measure $\Gamma_2^{(1)}$ at a liquid/vapour interface, although accurate measurements are possible for liquid/solid interfaces. In the next section we shall see how this quantity can be related to the surface tension of the interface and obtainable from experimental measurements of surface tension.

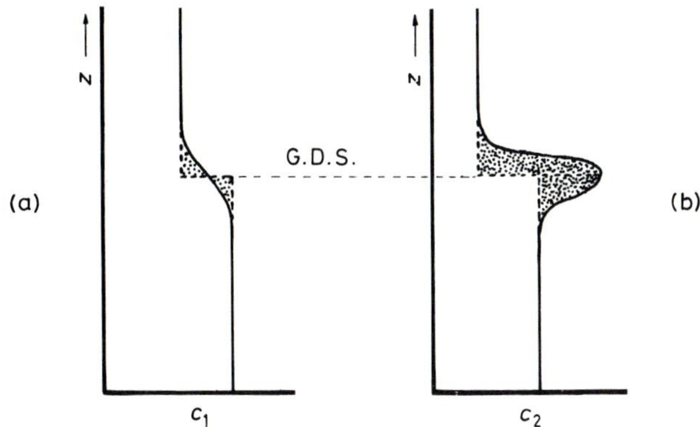

Figure 5.5 *Concentration profiles of (a) component 1 and (b) component 2 at an interface. If the Gibbs dividing surface is chosen so that the adsorption of component 1 is zero [i.e. the stippled areas in (a) are equal], then the stippled area in (b) gives the relative adsorption of 2 with respect to component 1:* $\Gamma_2^{(1)}$.

THE GIBBS ADSORPTION EQUATION

One of the most important equations in surface thermodynamics is that which links changes of surface tension to adsorption processes. The derivation of this equation is given by Appendix III.

The equation is most commonly applied to the case in which the temperature is constant, when it is known as the *Gibbs adsorption isotherm*, and has the form, for a binary mixture,

$$d\sigma = -\Gamma_2^{(1)} d\mu_2, \qquad (5.2)$$

μ_2 being the chemical potential of component 2. Since at equilibrium the chemical potential of 2 is uniform throughout the system, μ_2 may be taken as its value in either of the adjacent phases. To make use of this equation we therefore express this chemical potential in terms of the composition of one of the bulk phases. For example, for a liquid/vapour interface we may equate μ_2 to its value in the liquid:

$$d\mu_2 = RT d\ln a_2^l = RT d\ln x_2^l \gamma_2^l, \qquad (5.3)$$

where a_2^l is the activity of 2, set equal to the product of its mole fraction, x_2^l, and its activity coefficient, γ_2^l.

Hence,

$$d\sigma = -RT\Gamma_2^{(1)} d\ln x_2^l \gamma_2^l. \qquad (5.4)$$

Since in any stable phase the chemical potential increases with increase in concentration, it follows that if component 2 is adsorbed at the interface ($\Gamma_2^{(1)}$ positive) then the surface tension decreases with increase in concentration of component 2.

Substances that are strongly adsorbed at an interface and hence cause a substantial lowering of the surface tension are called *surface-active agents* or *surfactants*. The difference between the surface tension (σ) of a solution of a surface-active agent at a mole fraction of x_2^l and that of the pure solvent (σ_2^*) is given simply by integrating equation (5.2) or (5.4):

$$\sigma - \sigma_1^* = -\int_{x_2=0}^{x_2} \Gamma_2^{(1)} \mathrm{d}\mu_2 \qquad (5.5)$$

or

$$\sigma - \sigma_1^* = -RT \int_{x_2=0}^{x_2} \Gamma_2^{(1)} \mathrm{d}\ln x_2^l \gamma_2^l. \qquad (5.6)$$

An important application of the Gibbs adsorption equation is to the calculation of the relative adsorption from measurements of the variation of surface tension with concentration:

$$\Gamma_2^{(1)} = -\frac{1}{RT}\left[\frac{\partial \sigma}{\partial \ln x_2^l \gamma_2^l}\right]. \qquad (5.7)$$

In the special case of an ideal solution $\gamma_2^l = 1$ and $\mathrm{d}\ln x_2 = \mathrm{d}x_2/x_2$, so that

$$\Gamma_2^{(1)} = -\frac{x_2^l}{RT}\left[\frac{\partial \sigma}{\partial x_2^l}\right]. \qquad (5.8)$$

The early attempts to measure $\Gamma_2^{(1)}$ directly, outlined in Appendix III, were designed to test the Gibbs equation experimentally, by comparing the measured value with that calculated from surface tension data using equation (5.7). The limited accuracy of these measurements did no more than check this equation to within a few percent. However, since equation (5.7) is derived by rigorous thermodynamic arguments, this comparison should be regarded as assessing the reliability of the experimental techniques used for the direct measurement of adsorption and surface tension rather than a test of the correctness of the thermodynamic equation.

The Gibbs equation applies also to the solid/liquid interface. Direct measurements of the surface tension of a solid are possible only in very special circumstances. On the other hand, as outlined in Appendix III, adsorption at a solid surface is usually more readily measured. In this case equation (5.7) may be used to calculate the surface tension at the solid/solution interface, relative to that at the interface between solid and pure component 1. This contrast between fluid/fluid and fluid/solid interfaces is summarised in Table 5.1.

Table 5.1 *Measurement of surface tension at fluid/fluid and fluid/solid interfaces.*

	Fluid/fluid	*Fluid/solid*
σ	Directly measurable	Not measurable; calculate from adsorption data
Γ	Not accurately measurable; calculate from surface tension data	Directly measurable

THE INFLUENCE OF ADSORPTION ON INTERPARTICLE FORCES

We have identified the free-energy increase arising from the formation of unit area of new surface with the surface tension and have pointed out that when the surfaces are immersed in a fluid (liquid or vapour) this free energy is modified (*i.e.* the surface tension of the approaching surfaces is decreased). One contribution to this effect arises from the screening of London–van der Waals forces by the intervening medium. A second contribution arises from adsorption by the surfaces of one of the components of the fluid phase.

To fix our ideas let us consider two infinite, flat, parallel plates in adsorption equilibrium with a binary solution. Suppose that at a separation H and at a given solution concentration the relative adsorption of component 2 is $\Gamma_2^{(1)}(H)$. We wish to know how the presence of the solution between the plates at a mole fraction x_2 affects the surface tension.

According to the Gibbs equation,

$$d\sigma/d\mu_2 = -\Gamma_2^{(1)}(H). \tag{5.9}$$

Differentiating this equation with respect to H we obtain

$$d^2\sigma/d\mu_2 dH = -d\Gamma_2^{(1)}(H)/dH. \tag{5.10}$$

Since the order of differentiation does not matter, this can be written in the form

$$\mathrm{d}(\mathrm{d}\sigma/\mathrm{d}H)/\mathrm{d}\mu_2 = -\mathrm{d}\Gamma_2^{(1)}(H)/\mathrm{d}H. \tag{5.11}$$

But $2\mathrm{d}\sigma/\mathrm{d}H = -\mathcal{F}$, so that

$$\mathrm{d}\mathcal{F}(H)/\mathrm{d}\mu_2 = 2\mathrm{d}\Gamma_2^{(1)}(H)/\mathrm{d}H. \tag{5.12}$$

Recalling that a positive value of $\mathcal{F}$ implies a repulsion, we are led to the following important qualitative generalisations concerning the effect of adding component 2 to the solution:

(1) if the adsorption increases as the plates are brought together [*i.e.* if $\mathrm{d}\Gamma_2^{(1)}/\mathrm{d}H$ is negative for all values of μ_2], then the magnitude of the attractive force between the plates increases as component 2 is added, while

(2) if the adsorption decreases as the plates are brought together [*i.e.* if $\mathrm{d}\Gamma_2^{(1)}/\mathrm{d}H$ is positive for all values of μ_2], then adsorption effects will make a contribution to the repulsive forces. If this contribution outweighs the attraction arising from dispersion forces, then the surfaces will repel one another.

These effects can be given a simple qualitative molecular interpretation. As the two surfaces come together, their adsorption force fields begin to overlap and produce a larger (deeper) adsorption potential [Figure 5.6(b)]. The adsorption will increase as the plates approach and an extra attractive force arises. However, when the space between the plates reaches molecular dimensions, although the adsorption potential may be high, the space available for adsorption decreases and at some point $\Gamma_2^{(1)}$ begins to decrease [Figure 5.6(c)]; adsorption now contributes to a repulsion between the plates. One way of looking at this effect is to say that to decrease the adsorption the system must be supplied with the desorption energy of the molecules which are being 'squeezed out' of the space between the plates; this energy is provided by the work done in pushing the plates together against a repulsive force. The above principles turn out to be of wide applicability, but it must be remembered that they refer to the case in which adsorption equilibrium is maintained as the surfaces come together.

In many kinds of system, particularly those involving polymer adsorption, it is doubtful whether there is time for adsorption equilibrium to become established during a collision between colloidal particles undergoing Brownian motion. Under these circumstances the collision takes place at constant adsorption and the chemical potential of the adsorbed species can no longer be equated with that of the species in solution. Consequently, a different interpretation has to be placed on the phenomenon as explained in Chapter 3 and dealt with further in Chapter 9 and Appendix VI.

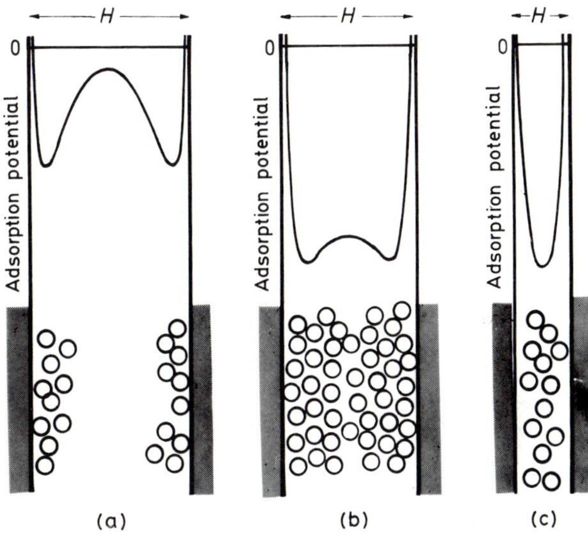

Figure 5.6 *Effect of adsorption on interparticle forces.* (a) *At large separations the adsorption at the two surfaces is independent of separation; adsorption does not contribute to the interparticle force.* (b) *At closer separations the adsorption fields of the two surfaces overlap and enhance the adsorption in the space between the surfaces:* $\partial\Gamma_2^{(1)}/\partial H$ *is negative so addition of component 2 leads to an attraction [*$\mathcal{F}(H)$ *negative].* (*c*) *At very close approach the space available for adsorption decreases:* $\partial\Gamma_2^{(1)}/\partial H$ *is positive and adsorption effects make a repulsive contribution to the interparticle force.*

This figure illustrates the effect in the case of adsorption from the vapour phase. For adsorption from solution where there is competition between the adsorption of solute and solvent the picture is slightly more complicated, but the outcome is similar.

THE EFFECT OF THE CURVATURE OF SURFACES ON EQUILIBRIUM

The concept of surface tension was introduced in Chapter 2 in terms of the amount of work needed to increase the area of a surface. In the case of liquids we may consider this to be done by forcing liquid reversibly out of a nozzle (Figure 5.7). Since this increases the total surface energy, work has to be done. If the piston is moved in displacing a volume $\mathrm{d}V$ against a pressure $(p^l - p^g)$, the work done is $(p^l - p^g)\mathrm{d}V$. The radius of the spherical drop (we ignore gravity for simplicity) increases from r to $r + \mathrm{d}r$, and its area increases by $8\pi r\mathrm{d}r$. The increase in energy for unit increase in surface area is equal to the surface tension:

$$\sigma = [(p^l - p^g)/8\pi r](\mathrm{d}V/\mathrm{d}r). \tag{5.13}$$

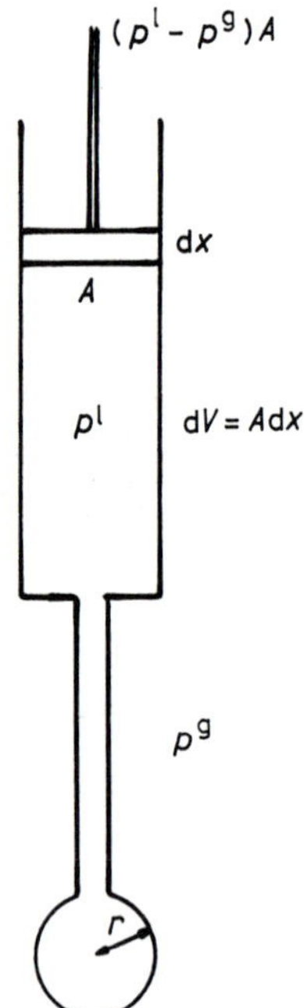

Figure 5.7 *Work done in increasing the area of a spherical drop. A force $(p^l - p^g)A$ has to be applied to the piston. The work done in moving the piston a distance dx is $(p^l - p^g)Adx$, which is equal to the work needed to increase the surface area of the drop: $8\pi\sigma r dr$. Note that the increase in the volume of the drop is equal to minus the decrease in the volume in the cylinder.*

Now for a sphere $\mathrm{d}V/\mathrm{d}r = 4\pi r^2$, and so

$$\sigma = (p^l - p^g)r/2$$

or

$$(p^l - p^g) = 2\sigma/r. \tag{5.14}$$

This is a special case of the *Laplace equation*, which can be expressed more generally as

$$(p^\alpha - p^\beta) = \Delta p = C\sigma, \tag{5.15}$$

where the phases α and β are separated by a surface whose curvature C is defined by

$$C = (1/r_1) + (1/r_2). \tag{5.16}$$

Here r_1 and r_2 are the principal radii of curvature of the surface (Figure 5.8). If the centre of curvature corresponding to r_1 lies in α, then r_1 is given a positive sign, and similarly for r_2. If r_1 and r_2 are both positive, then so is C and the pressure in α is higher than that in β. If r_1 and r_2 are of opposite sign [*e.g* for a saddle-shaped surface, Figure 5.8(b)], C may be positive or

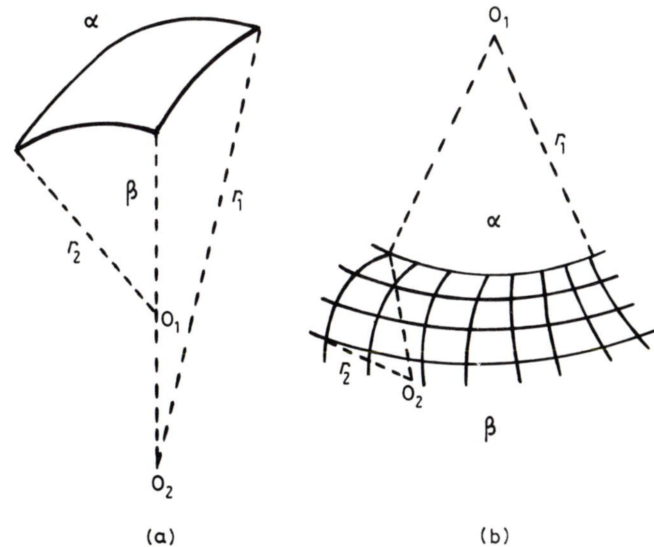

Figure 5.8 *Definition of the curvature C of a surface:* $C = (1/r_1) + (1/r_2)$. (a) *r_1 and r_2 of the same sign,* (b) *r_1 and r_2 of opposite sign.*

negative depending on the relative magnitudes of r_1 and r_2. If the pressure difference across the surface is the same at all points, then the surface must be one of constant curvature.

Since the properties of matter depend on pressure, the properties of the liquid in a drop will be different from those of the liquid in bulk. For liquids and solids the effects are negligible for drops of radius greater than a few micrometres; below this, however, important phenomena arise. Thus curvature effects become significant when one enters the colloid size range.

These effects are most simply seen by considering the dependence of the chemical potential on pressure,

$$\mathrm{d}\mu = v\mathrm{d}p, \tag{5.17}$$

where v is the molar volume. The chemical potential of liquid in a spherical drop is thus increased by

$$\mu^l\text{ (drop)} - \mu^l\text{ (bulk)} = 2\sigma v/r. \tag{5.18}$$

Now the chemical potential of a vapour (μ^g) is related to its partial pressure (p) by

$$\mu^g = \mu^{\dagger,g} + RT\ln(p/p^\dagger), \tag{5.19}$$

where $p^\dagger$ is a standard pressure (usually taken as the unit of pressure) at which $\mu = \mu^\dagger$. At equilibrium between vapour and bulk liquid, which exerts a vapour pressure p^*,

$$\mu^l\text{ (bulk)} = \mu^g(p^*) = \mu^{\dagger,g} + RT\ln(p^*/p^\dagger), \tag{5.20}$$

while for the liquid in a drop, at equilibrium with vapour at a pressure p,

$$\mu^l\text{ (drop)} = \mu^g(p) = \mu^{\dagger,g} + RT\ln(p/p^\dagger). \tag{5.21}$$

Inserting these values of μ^l (bulk) and μ^l (drop) into equation (5.18) gives

$$\ln(p/p^*) = 2\sigma v/rRT. \tag{5.22}$$

More generally,

$$\ln(p/p^*) = C\sigma v/RT. \tag{5.23}$$

Equations (5.22) and (5.23) are forms of the *Kelvin equation*.

Since for a drop C is positive, the vapour pressure of a drop is greater than that of the bulk liquid; conversely for liquid condensed in a capillary (Figure 5.9) C is negative and the vapour pressure is lowered. In this case C is related to the size of the capillary ($C = 2/r$ if the contact angle is zero). Measurements of the vapour pressure exerted by a liquid condensed in a porous solid can thus give information on the sizes of the pores present.

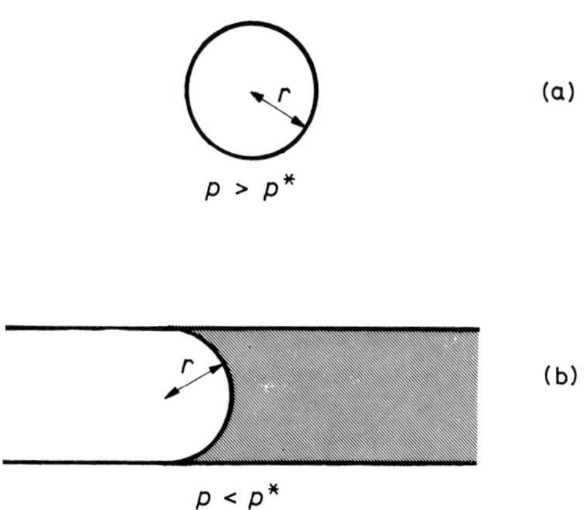

Figure 5.9 *Vapour pressure over curved surfaces:* (a) *over a drop* $p > p^*$, (b) *over liquid condensed in a cylindrical capillary* $p < p^*$.

Even though the surfaces bounding small crystals are not generally curved, they exhibit a similar increase in vapour pressure and in their solubility as their size is reduced. In this case the radius of curvature has to be replaced by a length (h), which for the present purpose, for an isodimensional crystal, can be taken as roughly half the diameter of the crystal. Thus

$$\ln(S/S^*) = 2\sigma v/hRT, \qquad (5.24)$$

where S is the solubility of the small crystal and S^* that of the bulk. Equation (5.24) has important consequences for colloidal particles, as will be mentioned in Chapter 10.

In the above it has been assumed that σ is independent of the drop or particle size. This may not always be justified and in more sophisticated theories this has to be taken into account.

Chapter 6

Some Important Properties of Colloids I Kinetic Properties

INTRODUCTION

In this chapter we shall discuss a group of properties related to the movement of colloidal particles. First we examine the problem of the random *Brownian motion* of a single particle and then apply the resulting equation to discuss the *diffusion* of particles down a concentration gradient. We then consider various *osmotic* phenomena. Next we deal with the diffusion of charged particles in an electric field, *electrophoresis*, and the related phenomena of *electro-osmosis* and *streaming potential*. Finally we treat briefly *sedimentation* and *creaming* in a gravitational field.

BROWNIAN MOTION

The apparently random stepwise or zig-zag movement of colloidal particles (Figure 6.1) was first observed by the botanist Robert Brown in 1827, and named after him. It provided early evidence for the molecular kinetic theory and was interpreted as arising from the random buffeting or jostling of the particles by molecules of the surrounding medium. The directions of movement of the molecules of the medium immediately adjacent to the particles are randomly oriented, while their speeds are distributed according to the Maxwell–Boltzmann law. The force acting upon the surface of a colloidal particle is proportional both to the frequency with which molecules collide with it and to the velocity of these molecules. The former is proportional to the local density of the molecules within one free path of the surface. Since the local density in a liquid fluctuates, as does the molecular velocity, the force on unit area of particle surface will also fluctuate. At any instant there will be a net force on a particle arising from the imbalance of forces on various parts of its surface [Figure 6.2(a)]. The direction and magnitude of the resultant force will vary randomly from instant to instant, leading to a zig-zag motion of the particle.

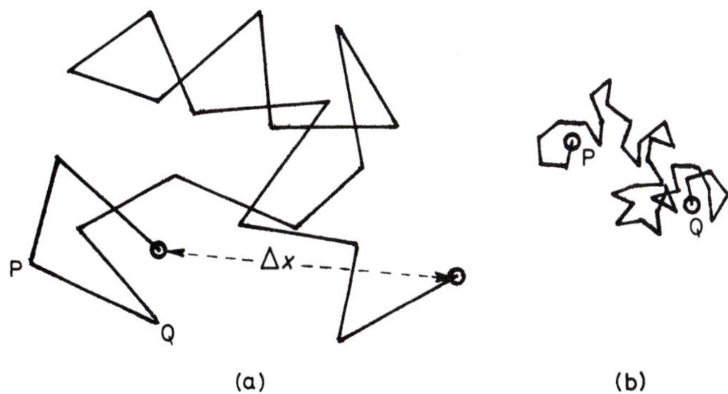

Figure 6.1 *Schematic representation of the random stepwise path executed by a particle undergoing Brownian motion.* (a) *The position of the particle is shown after successive time intervals* δt, *and the total displacement* Δx *after a time* Δt. (b) *If* δt *is long compared with the interval between collisions of molecules of the medium with the particle, then each of the steps in* (a), *e.g.* PQ, *is itself made up of a sequence of random substeps executed on a shorter timescale, as shown in* (b). *The shorter* δt *is made, the more fine-grained is the representation of the motion.*

One method of discussing this problem, due to Langevin (1908), is to consider the equation of motion of the particle as determined by the randomly varying instantaneous force acting upon it. The motion resulting from this force is opposed by the resistance arising from viscous forces, which is proportional to the velocity of the particle. Solution of this equation leads to an expression for the mean square velocity of a large number of particles (or the mean for a given particle averaged over a sufficiently long time interval). According to the *principle of equipartition of energy*, the mean square velocity is equal to $3kT/m$, where m is the mass of the particle. This can be shown to lead to the result

$$\overline{\Delta x^2} = (2kT/B)\Delta t, \tag{6.1}$$

where Δx is the distance travelled by a particle in the time Δt and B is the viscous *friction coefficient*. For spherical particles use can be made of Stokes' equation relating B to the viscosity of the medium (η) and the radius of the particle (a):

$$B = 6\pi\eta a, \tag{6.2}$$

whence

$$\overline{\Delta x^2} = (kT/3\pi\eta a)\Delta t. \tag{6.3}$$

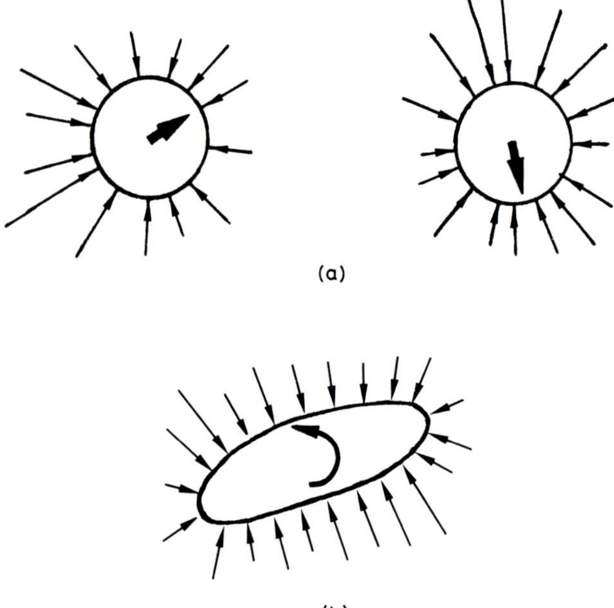

Figure 6.2 (a) *Resultant force on a particle arising from random forces caused by impacts with molecules of the medium; the direction and magnitude of the resultant force vary randomly from instant to instant.* (b) *Rotational Brownian motion of a non-spherical particle which is superimposed on the translational motion.*

Alternatively, one can regard the motion of the particle as a 'random walk', consisting of a sequence of steps, the direction of each step being chosen at random and independent of the direction of the preceding step. In reality the lengths of these steps, determined by fluctuations in the surrounding medium, will vary randomly, but it is sufficient to consider the case in which the steps are of equal length (λ). It can then be shown that the mean displacement of a particle after n steps is given by

$$\overline{\Delta x^2} = \lambda^2 n. \tag{6.4}$$

Now the number of steps taken in a time Δt is t/τ, where τ is the mean duration of a step, so that

$$\overline{\Delta x^2} = (\lambda^2/\tau)\Delta t. \tag{6.5}$$

Again appealing to the equipartition principle, it can been shown that

$$\lambda^2/\tau = 2kT/B, \tag{6.6}$$

leading back to equations (6.1) and (6.3).

Note that the mass of the particle does not appear explicitly in equation (6.3) but the mean square displacement is proportional to the reciprocal of its radius and hence to the reciprocal of the cube root of its volume. This means that the smaller the particle the more extensive the Brownian motion. We also observe the important result that the root mean square displacement is proportional to the square root of the time.

If the particles are non-spherical then, first, the simple form of Stokes' equation does not apply and, secondly, the unsymmetrical forces exerted on the particle by solvent molecules cause the particles to rotate and undergo Brownian rotation or *rotational diffusion* [Figure 6.2(b)].

DIFFUSION

Brownian motion is a characteristic of the movement of single colloidal particles, but this motion has important consequences when many particles are present. Suppose, for example, we consider a thin sheet in which there are initially c^0 particles in unit volume [Figure 6.3(a)] and examine the distribution of these particles after a time Δt. They will have spread out in both directions. The chance that a given particle will have reached a distance Δx is proportional to $\Delta t^{1/2}$: the sharp initial concentration peak will spread out into a broad peak, which has the shape of a Gaussian probability curve [Figure 6.3(b)].

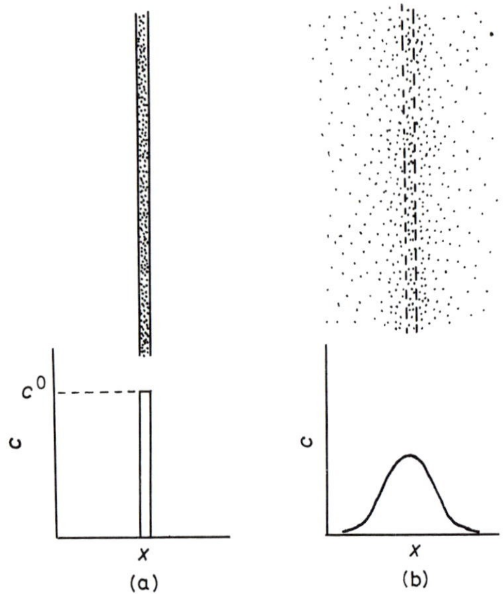

Figure 6.3 *Diffusion of particles from a thin sheet at a concentration c^0:* (a) *at zero time,* (b) *at a later time.*

The local concentration will now be proportional to the product of the initial concentration c^0 and a probability function whose width increases with $\Delta t^{1/2}$.

Now let us consider two sheets I and II a short distance apart, the concentration in one being c_1 and in the other c_2, where $c_1 > c_2$ (Figure 6.4), and examine the flow across some intermediate plane (shown dashed). Particles will move through this plane from both directions, but the flow from I will be greater than that from II so that there is a net flow from the higher to the lower concentration. This net flow per unit area will be equal to $(c_1 - c_2)u$, where u is the mean rate of flow of individual molecules along the x-axis.

From the theory of Brownian motion we already have an expression for $\overline{\Delta x^2}$ [equation (6.3)]. It is sufficient to make the approximation $\overline{\Delta x^2} \rightarrow (\overline{\Delta x})^2$, so that the mean speed is

$$u = \mathrm{d}\Delta x/\mathrm{d}t = kT/6\pi\eta a\Delta x. \tag{6.7}$$

The net rate of flow of particles is thus

$$\mathrm{d}n/\mathrm{d}t = u(c_1 - c_2) = -u(\mathrm{d}c/\mathrm{d}x)\Delta x$$

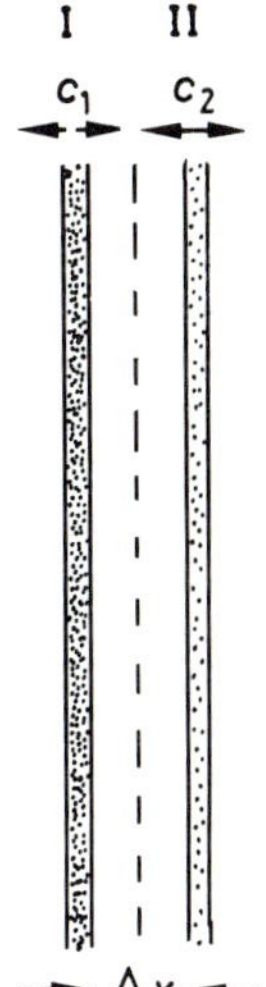

Figure 6.4 *Diffusion from two closely neighbouring sheets across a plane between them.* $c_1 > c_2$, $c_1 = c_2 - (dc/dx)\Delta x$.

or, from equation (6.7),

$$\mathrm{d}n/\mathrm{d}t = -[kT/(6\pi\eta a)](\mathrm{d}c/\mathrm{d}x). \tag{6.8}$$

Experimentally it is indeed found that the flow is directly proportional to the concentration gradient. This is *Fick's first law of diffusion*:

$$\mathrm{d}n/\mathrm{d}t = -D(\mathrm{d}c/\mathrm{d}x), \tag{6.9}$$

where D is called the *diffusion coefficient*. Thus

$$D = kT/6\pi\eta a. \tag{6.10}$$

It follows that another way of writing equation (6.3) for Brownian motion is

$$\overline{\Delta x^2} = 2D\Delta t. \tag{6.11}$$

The above interpretation of diffusion is based on a molecular model of Brownian motion and follows broadly the arguments of Einstein (1905).

We may also derive the same result from a thermodynamic argument. Just as a ball falls from a higher to lower gravitational potential and as heat flows from a higher to a lower temperature, so do molecules diffuse from a higher to lower chemical potential (μ). In general, the negative of the differential of a potential with respect to distance may be regarded as a force. Thus the gravitational potential of a mass m is $mg\Delta h$ and $-\mathrm{d}(mg\Delta h)/\mathrm{d}h = -mg$ is the gravitational force acting on the mass; similarly the 'force' driving heat flow is $-\mathrm{d}T/\mathrm{d}x$. By analogy, $-\mathrm{d}\mu/\mathrm{d}x$ is the 'chemical force' driving molecules to a lower potential. In the case of steady flow of particles this force is exactly balanced by the viscous resistance, which is proportional to the overall rate of flow u, so that

$$-\mathrm{d}\mu/\mathrm{d}x = Bu. \tag{6.12}$$

Now according to thermodynamic arguments, the chemical potential for a single particle is

$$\mu = \mu^{\ominus} + kT\ln c, \tag{6.13}$$

where $\mu^{\ominus}$ is a standard potential, independent of c and hence of x, so that

$$-\mathrm{d}\mu/\mathrm{d}x = -(kT/c)(\mathrm{d}c/\mathrm{d}x) = Bu. \tag{6.14}$$

Thus

$$u = -(kT/Bc)(\mathrm{d}c/\mathrm{d}x). \tag{6.15}$$

The amount of material transported per unit area is

$$\mathrm{d}n/\mathrm{d}t = uc = (kT/B)(\mathrm{d}c/\mathrm{d}x) \tag{6.16}$$

or

$$D = kT/B. \tag{6.17}$$

Again, for spherical particles B is given by equation (6.2) and

$$D = kT/6\pi\eta a. \tag{6.18}$$

OSMOSIS

The phenomenon of osmosis is another example in which molecules flow from a location of higher chemical potential to a region of lower potential. In an osmotic experiment a sample of dispersion in compartment II is separated from pure medium in compartment I by a semi-permeable membrane which allows passage of molecules of the medium but is completely impermeable to particles of the suspension [Figure 6.5(a)]. The chemical potential of the pure solvent ($\mu_1^{*\mathrm{I}}$) in I is higher than that of the solvent molecules in the dispersion in II (μ_1^{II}). For an ideal system

$$\mu_1 = \mu_1^* + RT\ln x_1$$

so that

$$\mu_1^{*\mathrm{I}} - \mu_1^{\mathrm{II}} = -RT\ln x_1 > 0, \tag{6.19}$$

where x_1 is the mole fraction of solvent in II. There will therefore be a spontaneous diffusion of solvent molecules into the suspension, which becomes progressively more dilute. If the pressure on the two compartments remains constant, then this process continues indefinitely until the suspension becomes infinitely dilute ($x_1 \rightarrow 1$).

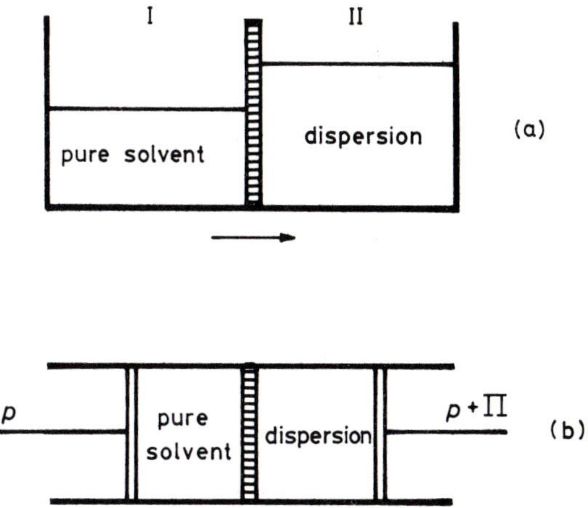

Figure 6.5 *Osmosis.* (a) *Dispersion separated from pure solvent by a semi-permeable membrane. Solvent molecules diffuse into the dispersion.* (b) *Osmotic equilibrium is set up when an excess pressure* Π *is applied to the dispersion to stop the flow of solvent through the membrane.*

The situation is changed, however, if the pressure applied to the dispersion in II is increased [Figure 6.5(b)]: the osmotic flow is slowed down and stops when the pressure difference between the two compartments reaches the *osmotic pressure* (Π) of the dispersion. At this point the excess pressure, Π, applied to the dispersion has raised the chemical potential of the solvent in compartment II, $\mu_1^{II}(p + \Pi)$, to equal that of the pure solvent in I, $\mu_1^{*I}(p)$.

The increase in chemical potential of the solvent when the applied pressure is increased by dp is $d\mu_1 = v_1 dp$, so that

$$\Delta\mu_1^{II} = v_1\Pi = \mu_1^{*II}(p + \Pi) - \mu_1^{*I}(p), \tag{6.20}$$

where v_1 is the molar volume of the solvent.

Therefore,

$$\mu_1^{II}(p + \Pi) = \mu_1^{*II}(p + \Pi) + RT\ln x_1 = \mu_1^{*I}(p) \tag{6.21}$$

or

$$\mu_1^{*II}(p + \Pi) - \mu_1^{*I}(p) = \Pi v_1 = -RT\ln x_1. \tag{6.22}$$

Thus

$$\Pi = -(RT/v_1)\ln x_1. \tag{6.23}$$

This equation can be simplified if x_1 is close to unity (dilute dispersion) since then

$$\ln x_1 = \ln(1 - x_2) \simeq -x_2, \tag{6.24}$$

and

$$\Pi = RTx_2/v_1. \tag{6.25}$$

When the mass concentration of particles, $\rho = m_2/V$, replaces the mole fraction, *i.e.**

$$x_2 = v_1\rho/M_2, \tag{6.26}$$

the equation becomes

$$\Pi = RT\rho/M_2, \tag{6.27}$$

which is *van't Hoff's equation*.

In principle, therefore, the molar mass of the particles can be calculated from measurements of Π as a function of ρ in a dilute suspension. In practice, the magnitude of Π is so small for particulate dispersions (*e.g.* for $c_2 = 10^{15}$ particles cm^{-3}, $\Pi = 4 \times 10^{-5}$ atm) that osmometry does not provide a useful route to particle size determination. It is nevertheless used for the determination of the molar mass of macromolecules, although

* In a dilute suspension $x_2 \simeq n_2/n_1 = (m_2/M_2)/(V/v_1) = \rho v_1/M_2$, where M_2 is the molar mass of the particles.

here the applicability is limited both by the small values of Π when M_2 is large and by the difficulty of preparing suitable membranes when M_2 is small (say smaller than a few thousand g mol^{-1}).

Because solutions of macromolecules exhibit deviations from ideality even at low concentrations, this must also be taken into account. Equation (6.23) for an ideal mixture may be written

$$\Pi v_1/RT = -\ln(1 - x_2) = x_2 + \tfrac{1}{2}x_2^2 + \ldots \qquad (6.28)$$

If the solution is non-ideal, then by analogy with the virial equations used to represent the properties of non-ideal gases we may write

$$\Pi v_1/RT = x_2 + B'x_2^2 \ldots, \qquad (6.29)$$

where $B' = \frac{1}{2}$ for an ideal solution. However, when $B' = 0$, the system follows the van't Hoff equation to much higher concentrations than it would for an ideal solution. Writing equation (6.29) in terms of mass concentration, we have

$$\Pi/RT = \rho/M_2 + B'(v_1/M_2^2)\rho^2 \ldots \qquad (6.30)$$

The term $B'(v_1/M_2^2)$ is called the *second osmotic virial coefficient** and is denoted by B. Then

$$\Pi/\rho = RT/M_2 + BRT\rho. \qquad (6.31)$$

The molar mass M_2 can thus be obtained by measuring Π as a function of ρ, plotting Π/ρ against ρ, and extrapolating the line so obtained to zero concentration.

DONNAN EQUILIBRIUM

A somewhat more complex situation arises when osmotic equilibrium is set up between two electrolyte solutions, one of which contains ions to which the membrane is impermeable. These ions may be macromolecular (colloidal electrolytes) or in the limit

* This term is sometimes applied to B'. It is important to remember that the analogy between the second osmotic virial coefficient, B, and the second gas virial coefficient, B_g, is not exact. For an ideal gas $B_g = 0$, while for an ideal solution $B' = \frac{1}{2}$ and $B = \frac{1}{2}(v_1/M_2^2) \neq 0$. *N.B.* The osmotic virial coefficient is not to be confused with the osmotic coefficient defined as $\Pi/\Pi^{\text{ideal}} = \phi = 1 + (B' - \frac{1}{2})(v_1/M_2)\rho + \ldots$; for an ideal solution $\phi \rightarrow 1$ as $\rho \rightarrow 0$.

charged colloid particles. Under these circumstances equilibrium involves not only the establishment of a pressure difference but also an electrical potential difference across the membrane. To illustrate the principles underlying this phenomenon it is convenient to discuss a simple case and to assume that the mobile ions behave as ideal solutes.

We consider an initial state in which [Figure 6.6(a)] compartment I contains a solution of sodium chloride at a mole fraction x^{I} in contact through a membrane with a solution in compartment II of the sodium salt of a colloidal electrolyte, NaX. The anion X^- is unable to diffuse through the membrane. For convenience we assume that the two compartments contain the same amounts of material. Since the concentration of chloride ions in I is greater than that in II, chloride ions will flow from compartment I to II. To maintain electrical neutrality an equal amount of sodium ions must accompany them. Furthermore, diffusion of solvent will also take place to establish osmotic equilibrium. If diffusion lowers the salt concentration in compartment I by Δx, the resulting concentrations in the two compartments will have the values shown in Figure 6.6(b).

Equilibrium is achieved when the chemical potentials of *all* diffusable species are the same on the two sides of the membrane. Thus for the solvent, S,

$$\mu_{\mathrm{S}}^{\mathrm{I}} = \mu_{\mathrm{S}}^{\mathrm{II}}, \tag{6.32}$$

which requires the establishment of an osmotic pressure, Π, across the membrane given by [*cf.* equation (6.23)]:

$$\Pi = p^{\mathrm{II}} - p^{\mathrm{I}} = -(RT/v_{\mathrm{S}})\ln(x_{\mathrm{S}}^{\mathrm{II}}/x_{\mathrm{S}}^{\mathrm{I}}). \tag{6.33}$$

[In the previous section, equation (6.23) refers to the case in which compartment I contains pure solvent so that $x_{\mathrm{S}}^{\mathrm{I}} = 1$.]

For the diffusable ions

$$\begin{aligned} \mu_{\mathrm{Na}}^{\mathrm{I}} &= \mu_{\mathrm{Na}}^{\mathrm{II}}, \\ \mu_{\mathrm{Cl}}^{\mathrm{I}} &= \mu_{\mathrm{Cl}}^{\mathrm{II}}. \end{aligned} \tag{6.34}$$

Since the ionic concentrations in the two compartments are different, these equalities can only be achieved if other factors which influence the chemical potentials differ in the two compartments. The pressure difference makes some contribution which we shall neglect. The main effect arises from the electrical potential difference. It follows from the discussion in Chapter 3, page 40, that the chemical potential of an ionic species (now its *electrochemical potential*, $\tilde{\mu}$) depends on the electrical potential of the phase. For an ion of valency z a change in electrical potential $\Delta\psi$ causes a change in the electrochemical potential

$$\Delta \tilde{\mu} = zF\Delta\psi, \tag{6.35}$$

where F is Faraday's constant (96 487 coulombs mol^{-1}).

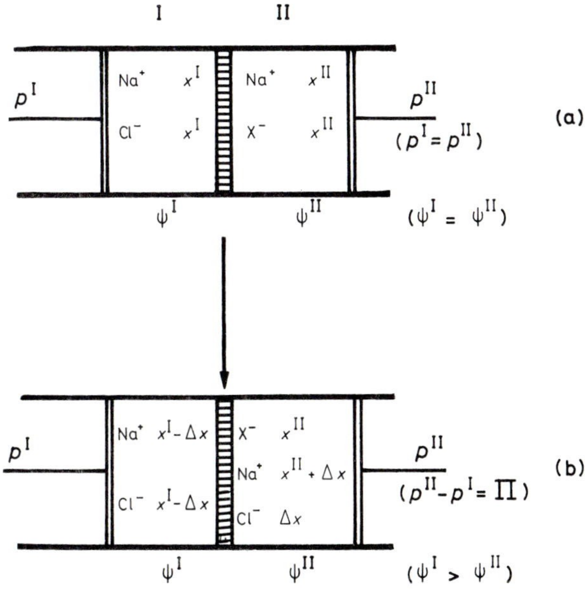

Figure 6.6 *The Donnan effect.* (a) *Initial state: the pressures and electric potentials are the same in the two compartments.* (b) *When equilibrium has been set up, an amount* Δx *of NaCl has diffused from compartment* I *to compartment* II *and a pressure difference,* Π, *and an electric potential difference* $\psi^{II} - \psi^{I}$ *have been established.*

This has to be added to the change in chemical potential arising from the concentration difference between the compartments, so that at equilibrium for sodium ions ($z = +1$)

$$\mu^{II}_{Na} - \mu^{I}_{Na} = RT\ln(x^{II}_{Na}/x^{I}_{Na}) + F(\psi^{II} - \psi^{I}) = 0. \tag{6.36}$$

Similarly, for chloride ions ($z = -1$)

$$\mu^{II}_{Cl} - \mu^{I}_{Cl} = RT\ln(x^{II}_{Cl}/x^{I}_{Cl}) - F(\psi^{II} - \psi^{I}) = 0. \tag{6.37}$$

Thus

$$\psi^{\mathrm{II}} - \psi^{\mathrm{I}} = (RT/F)\ln(x^{\mathrm{I}}_{\mathrm{Na}}/x^{\mathrm{II}}_{\mathrm{Na}}) = RT/F\ln(x^{\mathrm{II}}_{\mathrm{Cl}}/x^{\mathrm{I}}_{\mathrm{Cl}}), \quad (6.38)$$

whence*

$$x^{\mathrm{I}}_{\mathrm{Na}}x^{\mathrm{I}}_{\mathrm{Cl}} = x^{\mathrm{II}}_{\mathrm{Na}}x^{\mathrm{II}}_{\mathrm{Cl}}. \quad (6.39)$$

Inserting the concentrations given in Figure 6.6(b), we obtain for the amount of salt transferred

$$\Delta x = (x^{\mathrm{I}})^2/(x^{\mathrm{II}} + 2x^{\mathrm{I}}), \quad (6.40)$$

which is always positive.

Thus three effects are observed:

(1) An osmotic pressure is set up.
(2) There is a net transfer of salt into the compartment containing the non-diffusable ion. This applies whether this ion is an anion or a cation.
(3) An electrical potential difference (*membrane potential*) is set up across the membrane. In the present example $\psi^{\mathrm{II}} - \psi^{\mathrm{I}}$ is negative. If the non-diffusing ion had been a cation, $\psi^{\mathrm{II}} - \psi^{\mathrm{I}}$ would have been positive.

It is not possible here to investigate more complicated examples of the Donnan effect. Suffice it to say that membrane potentials (which can be produced by mechanisms other than the Donnan effect) play an important role in many colloid phenomena, especially in biological systems.

DIALYSIS

One practical application of osmosis is the process of *dialysis* in which colloidal dispersions are purified by removing from them traces of reagents or by-products remaining from, or produced in, their preparation. The colloid is contained in a bag of semi-permeable material, *e.g.* a Visking (cellulose) membrane. This is immersed in a bath of distilled water, which is replaced from time to time. The unwanted components, to which the membrane is permeable, diffuse out of the dialysis bag, and completion of the

* This result follows also from the fact that the chemical potential of the neutral species NaCl is $\mu_{\mathrm{NaCl}} = \mu_{\mathrm{Na}} + \mu_{\mathrm{Cl}}$ and its activity is $x_{\mathrm{Na}}x_{\mathrm{Cl}}$. Since the chemical potential of a neutral species is unaffected by the electrical potential, the condition of equilibrium is just equation (6.39).

process is monitored by measuring the conductivity and/or the surface tension of the external water.

ELECTROPHORESIS, ELECTRO-OSMOSIS, AND STREAMING POTENTIALS

Electrically charged particles tend to move under the influence of an electric field. Once again as a first approach to this problem we may set the electrical force acting on a particle carrying unit charge equal to the negative of the differential of the electrical potential, ψ. If the particle carries a charge q, then the force is

$$\mathscr{F}_{\mathrm{el}} = -q(\mathrm{d}\psi/\mathrm{d}x) = qE, \tag{6.41}$$

where E is the electric field strength. For a spherical particle the viscous force opposing this motion is

$$\mathscr{F}_{\mathrm{visc}} = Bu = 6\pi\eta a u, \tag{6.42}$$

where u is the speed of the particle [*cf.* equation (6.2)]. At this point it should be noted that a particle moving through the fluid carries with it a thin boundary layer of stagnant fluid, so that a in the above equation refers to the radius of the particle plus the thickness of this layer. For many purposes this difference is unimportant (which is why we have not drawn attention to it earlier), but in the present context it cannot be ignored.

When the particle is in a state of steady motion, the two opposing forces are equal so that

$$u = qE/6\pi\eta a. \tag{6.43}$$

Now the electrical potential (ζ) at the surface of a sphere of radius a and carrying a charge q is*

$$\zeta = q/4\pi\varepsilon a, \tag{6.44}$$

where ε is the permittivity of the medium in which it is immersed. In terms of the *surface potential* or *zeta-potential* (ζ)

$$u = \tfrac{2}{3}\varepsilon\zeta E/\eta. \tag{6.45}$$

* It is a result of electrostatic theory that a sphere carrying a charge q uniformly distributed over its surface behaves as though the charge is concentrated at the centre of the sphere. The electrical potential immediately outside the surface of the sphere is thus given by equation (6.44) (*cf.* page 39).

The zeta-potential of the particle refers to the potential at the outer limit of the boundary layer, often called the *shear plane* or *slipping plane*. It may be determined from measurement of the velocity of particles in an electric field E. Since the precise location of the shear plane is difficult to define, the zeta-potential is an ambiguous measure of the potential at the surface of the particle (Figure 3.5).

In the above argument we have ignored the effect of the ionic atmosphere, or electrical double layer, around the charged particle. Its presence has two consequences (Figure 6.7). First, the counter-ions in the double layer tend to move in a direction opposite to that of the particle, which in effect has to drag its double layer with it. Secondly, it does not entirely succeed in taking all of its double layer, but fresh ions become attached to it as some are left behind; this reconstruction of the double layer does not take place instantaneously. Both effects lead to a retardation of the movement of the particle. More complete theories lead to the introduction of an additional term $f(\kappa a)$ into equation (6.45):

$$u = \tfrac{2}{3}(\varepsilon\zeta E/\eta)\cdot f(\kappa a), \qquad (6.46)$$

where κ is the reciprocal Debye length introduced in Chapter 3.

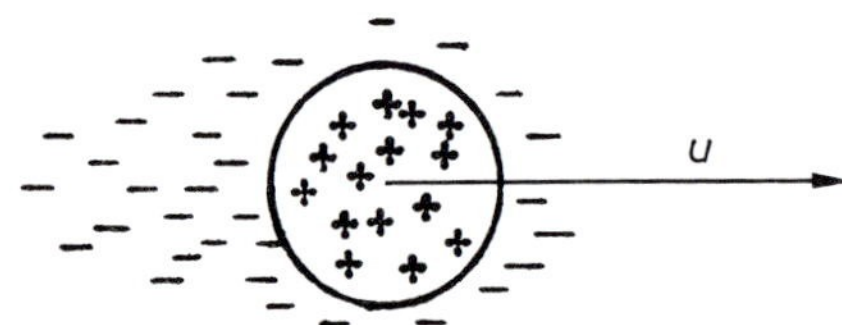

Figure 6.7 *Double layer around a moving particle in an electric field.*

For very small particles in dilute solution where the radius of the double layer ($1/\kappa$) is large so that $\kappa a \rightarrow 0$, the correction factor $f(\kappa a) \rightarrow 1.0$ and equation (6.45) still applies. However, for large particles in more concentrated solution where $\kappa a \rightarrow \infty$ (in practice > 100), $f(\kappa a) \rightarrow 1.5$. If the particles conduct electricity, a further correction is needed.

In addition to the above theoretical problems in interpreting electrophoretic data there is also an important experimental problem arising from a second electrokinetic effect, which we now consider.

In discussing electrophoresis we have considered the motion of a charged particle relative to the surrounding liquid. A similar phenomenon occurs when liquid in a capillary, whose walls carry an electric charge, is subjected to an electric field directed along the axis of the capillary. Since the solid is immobile it is the liquid which now moves relative to the solid surface. This is the phenomenon of *electro-osmosis*, which arises from the influence of the electric field on the counter-ions in the double layer adjacent to the wall [Figure 6.8(a)]. It turns out that, since the radius of the capillary is large compared with $1/\kappa$, the relative velocity of the liquid is given by equation (6.46) with $f(\kappa a) = 1.5$, *i.e.*

$$u = \varepsilon\zeta E/\eta. \tag{6.47}$$

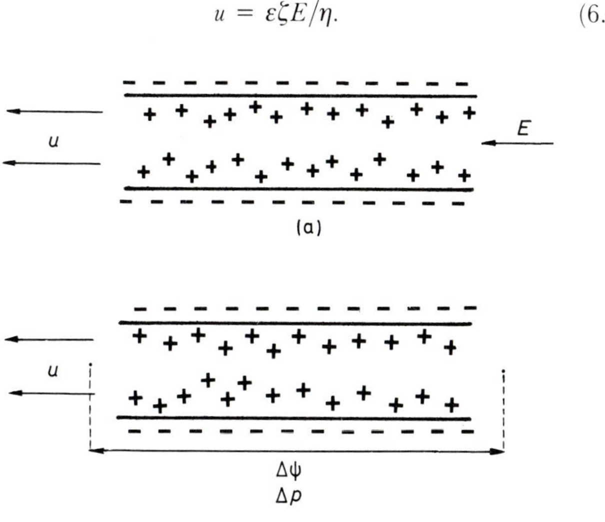

Figure 6.8 *Electrokinetic phenomena:* (a) *electro-osmotic flow of the medium caused by the flow of counter-ions under the influence of an electric field,* (b) *streaming potential set up when the liquid flows under the influence of a pressure gradient and carries counter-ions with it.*

Experimentally it is usual to measure the volume flow of liquid, $\mathrm{d}V/\mathrm{d}t$, which is given by

$$\mathrm{d}V/\mathrm{d}t = uA, \tag{6.48}$$

where A is the area of the capillary. For a cylindrical capillary this area is readily measured, but very often electro-osmosis is applied to the study of materials in the form of porous plugs whose area cannot be measured directly. It may now be assumed that Ohm's law can be applied. In this case the area and field strength can be related to the current, I, flowing through the plug:

$$I = AkE, \tag{6.49}$$

where k is the electrical conductivity of the solution. It follows that equations (6.47) and (6.48) can now be written

$$\mathrm{d}V/\mathrm{d}t = \varepsilon\zeta I/\eta k. \tag{6.50}$$

Again further corrections have to be made if the solid or its surface conducts electricity.

The direct observation of electrophoresis is carried out by observing the rate of movement of individual particles in a small cell or capillary illuminated in an ultramicroscope.* The walls of the container (glass or quartz) usually carry a surface charge so that the application of the electric field needed to observe electrophoresis causes the liquid adjacent to the walls to undergo electro-osmotic flow. Colloidal particles will be subject to this flow superimposed on their electrophoretic mobility. However, in a closed system the flow along the walls must be compensated for by a reverse flow down the centre of the capillary: a horizontal convection current is set up (Figure 6.9). There is a surface separating the two zones of convection at which the flow is zero, and it is necessary to make electrophoretic measurements only on particles at this level by adjusting the focus of the microscope to pick out such particles.

Finally, a third consequence of the electrokinetic effect is observed when an electrolyte is caused to flow through a capillary or porous plug by the application of a pressure difference, Δp. An electrical potential difference $\Delta\phi$, called the *streaming potential*, arising from the displacement of the double layer by the flowing liquid [Figure 6.8(b)], is set up. Its value is found to be given by

$$\Delta\phi = (\varepsilon\zeta/\eta k)\Delta p. \tag{6.51}$$

Comparing equations (6.50) and (6.51) we observe that

* The optics of an ultramicroscope produce a convergent beam of light that illuminates colloidal particles, which appear as bright specks on a dark field.

$$\Delta\phi/\Delta p = (\mathrm{d}V/\mathrm{d}t)/I = \varepsilon\zeta/\eta k, \tag{6.52}$$

showing the close relationship between electro-osmotic flow and streaming potential.

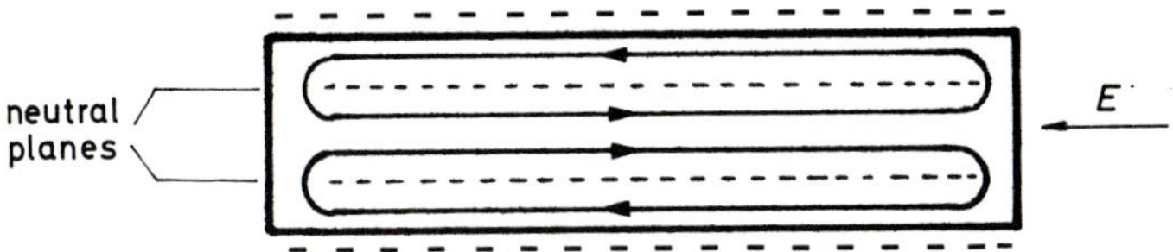

Figure 6.9 *Horizontal convective flow in an electrophoresis cell; observations must be made at the level of one of the neutral planes.*

SEDIMENTATION AND CREAMING

If the particles of a colloidal dispersion have a density different from that of the medium, they will be subjected to a gravitational force which will tend to cause the particles to sink (*sediment*), if they are denser than the medium, or rise to the surface (*cream*), if they are less dense. In either case the result will be to produce a concentration gradient of particles tending to restore a uniform concentration. Gravitational forces are thus opposed by diffusion, and a steady state is set up in which the two processes are in balance. As before, we can express the gravitational force as $-(\rho_\mathrm{p} - \rho_\mathrm{m})vg$, where ρ_p and ρ_m are the densities of particle and medium, respectively, and v is the volume of a particle. The driving forcc for diffusion is $(\mathrm{d}\mu/\mathrm{d}h)$. In the steady state

$$-(\rho_\mathrm{p} - \rho_\mathrm{m})vg = (\mathrm{d}\mu/\mathrm{d}h) = kT(\mathrm{dln}c/\mathrm{d}h), \tag{6.53}$$

so that

$$kT\mathrm{dln}c = -(\rho_\mathrm{p} - \rho_\mathrm{m})mg\mathrm{d}h. \tag{6.54}$$

Integrating this equation between a height h^0 and $h^0 + \Delta h$ we obtain

$$c/c^0 = \exp[-(\rho_\mathrm{p} - \rho_\mathrm{m})vg\Delta h/kT], \tag{6.55}$$

which is of course the result that we should expect from the application of the Boltzmann distribution law to the problem [equation (3.17)]. The above argument, however, brings out very clearly the way in which physico-chemical equilibrium is achieved by the balance between the effects of potential energy (gravitational energy here) and thermal energy (represented here by diffusion or Brownian motion).

The experimental test of equation (6.55), and also of (6.3), was first carried out by Perrin (1908) in his classical work on the sedimentation of gamboge particles in the Earth's gravitational field. This relationship is now widely applied in two main areas. It may be used, first, to determine particle sizes of very small particles and, secondly, by increasing the gravitational force using a centrifuge (or ultracentrifuge), to study both the diffusion and molar mass of macromolecules.

It is interesting to note that, although Boltzmann's distribution law was introduced to describe the statistical behaviour of molecules, the first experimental demonstration of its validity involved the assumption that colloidal particles follow the same statistical law as molecules. Perrin's experiments, although described in most text books as an early attempt to determine Avogadro's constant (N_A), should more logically be regarded as determining the Boltzmann constant, k, the evaluation of N_A being arrived at from the relation between k and the gas constant R:

$$R = N_A k. \tag{6.56}$$

As we shall see, the idea that colloidal systems can be treated by statistical mechanics in just the same way as molecular systems has been revived in recent years, and it represents one of the important growing points in colloid science.

Chapter 7

Some Important Properties of Colloids II Scattering of Radiation

INTRODUCTION

In this chapter we shall outline some of the properties of colloidal dispersions which can be studied by methods involving the interaction of radiation with matter. Classical methods based on the scattering of light have in recent years been supplemented by laser light scattering and neutron scattering techniques that are becoming of increasing importance.

LIGHT SCATTERING

Conventional Light Scattering

The first attempts at a systematic study of the optical properties of finely dispersed systems were those of Faraday. His preparation of colloidal gold was probably somewhat fortuitous, for the original objective of his work was to prepare very thin films of gold and to examine them by reflected and transmitted light. In the event the preparative method he tried led to colloidal gold. He regarded the scattering of a narrow beam of light by his preparations as evidence for the particulate nature of the gold, no doubt by analogy with the way in which dust in the atmosphere is revealed by a shaft of sunlight. Further study by Tyndall led to this phenomenon being called the *Tyndall effect*.

The theoretical study of the scattering of light by particulate matter also has a long history.* The first rigorous theory for very small particles was developed by Rayleigh in 1871. The landmarks in the further evolution of the subject were the extensions of the theory by Mie (1908), Debye (1915), and Gans (1925) to particles whose size approaches the wavelength of light. These led to methods of estimating the size, and in some circumstances the shape, of colloidal particles. Much further progress has been made in the last few years largely as a result of the development of lasers, of sophisticated electronics for recording data, and of computers both for the analysis of experimental results and for enabling theoretical calculations of complex mathematical functions to be carried out rapidly for comparison with experiment.

The subject is still developing rapidly and leading to new methods for the study of particle size and shape, rates of diffusion and particle growth, and the structure of dispersions. A detailed mathematical treatment would be out of place here, but nevertheless the basic principles can be expounded in a relatively straightforward fashion.

According to classical electromagnetic theory, radiation can be represented as a wave motion in which a fluctuating electrical field and, at right angles, an associated fluctuating magnetic field are propagated through space at the velocity of light. The radiation intensity (I) is proportional to the square of the amplitude (E^0) of the wave motion (Figure 7.1). We shall not be concerned here with the quantum mechanical interpretation in which the radiation is propagated as 'wave packets' or photons.

Radiation interacts with matter through the effects of the electric field vector on the electron distributions in molecules. *Absorption* of radiation involves raising a system from one energy level to a higher level by the absorption of a quantum of energy (a photon). *Elastic scattering* of radiation involves no such quantum jumps and can be discussed in classical terms.**

The experimental method of studying light scattering is, in principle, very simple. The basic arrangement is shown in Figure 7.2, in which the intensity of light scattered at an angle θ is

* The interested reader will find a fascinating account of the intertwining of work by many persons in M. Kerker, 'The Scattering of Light', Academic Press, 1967, pp. 54–63.

** Scattering may also be inelastic. The wavelength of the scattered light is then different from that of the incident light (Raman effect).

measured at a distance *r* from the scattering object by a suitable detector. In practice, the modern equipment to achieve high precision is sophisticated, and great care has to be taken both to exclude rigorously all traces of dust and to ensure that no stray light gains access to the detector.

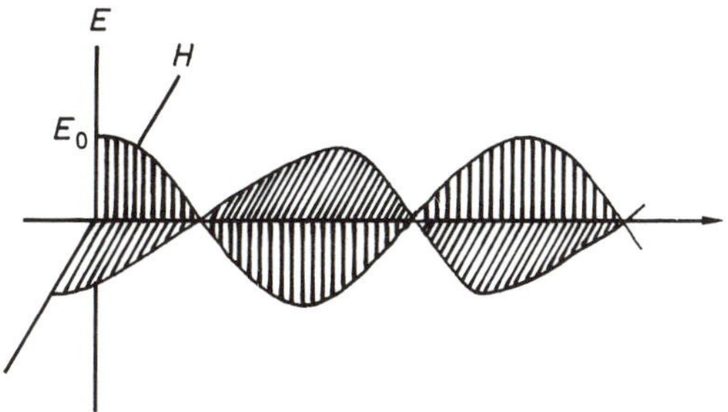

Figure 7.1 *Schematic representation of electromagnetic radiation as the propagation of a fluctuating electric field (E) at right angles to an associated fluctuating magnetic field (H). Light scattering results from the interaction of molecules with the electric vector.*

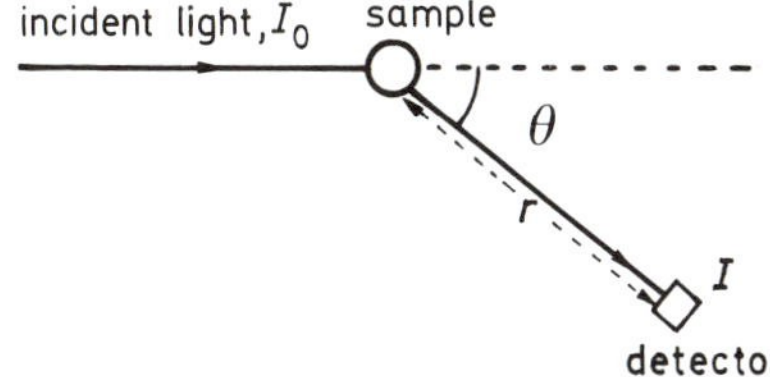

Figure 7.2 *Basic principle of the experimental arrangement for studying light scattering. Observations are made in the plane of the paper at an angle of θ to the direction of propagation.*

To understand the basic principles of light scattering it is best to begin with scattering from small particles whose dimensions are much smaller (say by a factor of twenty) than the wavelength of light: this corresponds to particles in the lower range of the colloidal size range (say less than 25 nm in diameter). At a given instant the particle will be subjected to an electric field which will polarise the particle, inducing in it a dipole moment (Figure 7.3). Since the particle is small, all parts of it are subjected to essentially the same field. As the wave passes, both the electric field and the induced dipole fluctuate in magnitude. If the light is plane polarised (*i.e.* the plane including the electric vector and the direction of propagation is fixed), the induced dipole lies in the

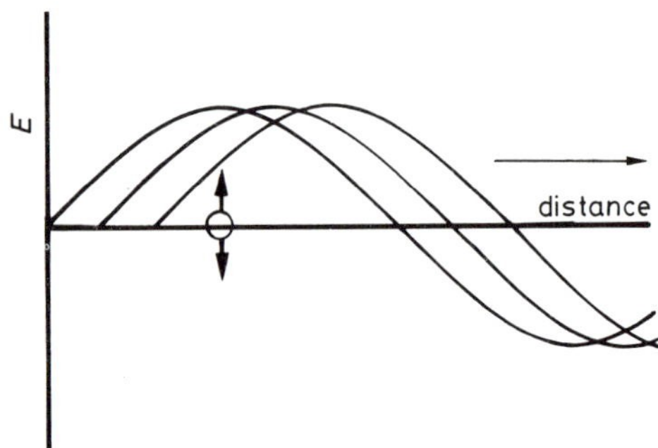

Figure 7.3 *Polarisation of a particle much smaller than the wavelength of light in the plane of the electric vector. The induced dipole is proportional to the amplitude of the wave at the position of the particle. As the wave passes the particle, the induced dipole oscillates with the same frequency as the light.*

plane of the electric vector, and part of the energy of the wave is used in creating the oscillating dipole. Since atomic nuclei are so much heavier than the electron, this dipole may be pictured as arising from the movement of electrons up and down in the plane of the electric vector. This electron movement constitutes in effect an oscillating current which produces its own electromagnetic field and radiates light of the same frequency as the incident radiation. The spacial distribution of the intensity of this radiation is sketched in Figure 7.4. The intensity of light propagated in a given direction is proportional to the length of the line, along the direction in question, from the origin to the surface depicted (the heavy arrows in Figure 7.4). Since the dipole has an axis of symmetry, the radiation intensity is symmetrical about this axis. This means that, if observations are made in the plane normal to the direction of polarisation of the light, the intensity at a given distance from the particle will be independent of the scattering angle, θ; this is indicated by the dashed line in Figure 7.4(a). Moreover, the light will be polarised in the same direction as the incident beam. Conversely, if the direction of polarisation is turned through 90° and observations are made in the same plane as before, so that this plane now includes the direction of polarisation [Figure 7.4(b)], then the radiation intensity is represented by the dumb-bell shape which results from the intersection of the plane of observation with the surface representing the spacial distribution (only one half of the surface is shown).

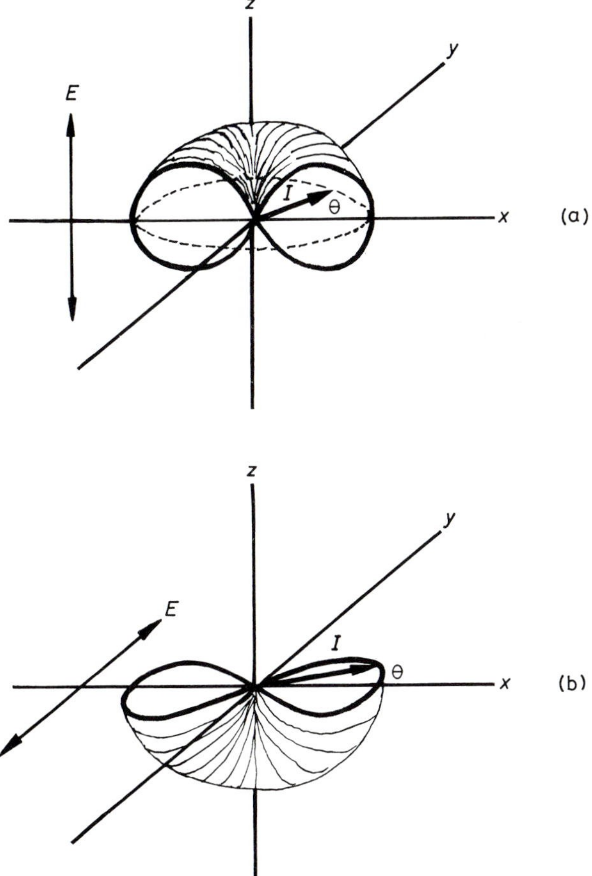

Figure 7.4 *Intensity of radiation from an oscillating dipole in space. Propagation is directed in the x-direction.* (a) *Electric vector in the x–z plane, observations made in the x–y plane. The intensity at an angle θ to the direction of propagation is given by the length of the heavy arrow and is independent of θ (dashed circle).* (b) *Electric vector in the x–y plane and observations made in the same plane. The intensity at an angle θ is given by the length of the heavy arrow and varies with θ, as shown by the heavy two-lobed curve.*

The intensity of a scattered wave at a distance r from its source is proportional to the square of the polarisability of the particle and inversely proportional to r^2. According to electromagnetic theory, the ratio of the scattered intensity (I) to the incident intensity (I_0) in the plane normal to the direction of polarisation

[Figure 7.4(a)] is given by

$$\left(\frac{I}{I_0}\right)_v = \frac{16\pi^4}{r^2\lambda^4}\left(\frac{\alpha}{4\pi\varepsilon_0}\right)^2, \tag{7.1}$$

where α is the polarisability of the particle, ε_0 the permittivity of a vacuum, and λ the wavelength of the light in the medium in which the particle is immersed. The subscript v denotes the fact that this refers to vertical polarisation with respect to the observational plane. Furthermore, the polarisability of a particle of volume v, having a refractive index n_1, immersed in a medium of refractive index n_0 is

$$\alpha = 3\varepsilon_0\left(\frac{n^2 - 1}{n^2 + 2}\right)v, \tag{7.2}$$

where $n = n_1/n_0$ is the *relative refractive index*.

Substitution for α in equation (7.1) then gives

$$\left(\frac{I}{I_0}\right)_v = \frac{9\pi^2}{r^2}\cdot\frac{1}{\lambda^4}\left(\frac{n^2 - 1}{n^2 + 2}\right)^2 v^2 \tag{7.3}$$

or, for a spherical particle of radius R, for which

$$v = (4\pi/3)R^3, \tag{7.4}$$

$$\left(\frac{I}{I_0}\right)_v = \frac{16\pi^4}{r^2}\cdot\frac{R^6}{\lambda^4}\left(\frac{n^2 - 1}{n^2 + 2}\right)^2. \tag{7.5}$$

Similarly, the intensity of light from a horizontally polarised beam [Figure 7.4(b)] is given by

$$\left(\frac{I}{I_0}\right)_h = \frac{16\pi^4}{r^2}\cdot\frac{R^6}{\lambda^4}\left(\frac{n^2 - 1}{n^2 + 2}\right)^2 \cos^2\theta. \tag{7.6}$$

Unpolarised light of intensity I_0 can be represented by the sum of two mutually perpendicular plane polarised beams of equal intensity $(I_0/2)$. Consequently, by combining the effects shown in Figures 7.4(a) and (b), the light scattered from unpolarised light and observed in the horizontal plane is the sum of contributions from the vertically and horizontally polarised components:

$$\left(\frac{I}{I_0}\right) = \frac{1}{r^2}\left[\frac{16\pi^4}{2}\cdot\frac{R^6}{\lambda^4}\left(\frac{n^2 - 1}{n^2 + 2}\right)^2(1 + \cos^2\theta)\right]. \tag{7.7}$$

The factor in square brackets is called the Rayleigh ratio, R_θ. The radiation diagram corresponding to equation (7.7) is shown in Figure 7.5. We see that when $\theta = 0°$ and $180°$ (forward and back scattering) $\cos^2\theta = 1$ and equal contributions are made by the vertical and horizontal components; the scattered light is unpolarised. When $\theta = 90°$, $\cos^2\theta = 0$ and only the vertical component contributes; the scattered light at this angle is vertically polarised.

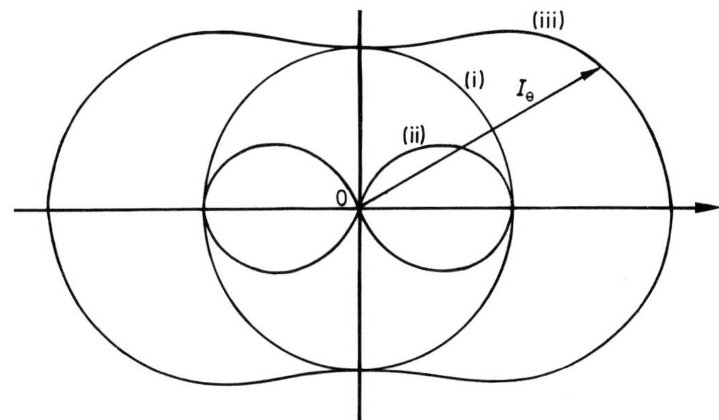

Figure 7.5 *Intensity of light scattered from an unpolarised incident beam by a small spherical particle at the origin as a function of the scattering angle (Rayleigh theory). The outer dumb-bell-shaped curve* (iii) *is the vectorial sum of the two inner curves* (i) *and* (ii), *given, respectively, by equations (7.5) and (7.6). The intensity at an angle θ is given by the length of the vector from the origin to the outer curve. The three-dimensional distribution is obtained by rotating the diagram about the long axis.*

The above calculations refer to light scattered by a single particle. Provided that a colloidal dispersion is not too concentrated, the scattering is simply the sum of contributions from individual particles. Thus the intensity of light scattered at an angle θ by unit volume of a dilute suspension of particle concentration c is

$$I/I_0 = R_\theta c/r^2. \tag{7.8}$$

The intensity of scattering is seen to depend, through R_θ, on the relative refractive index, and it falls to zero if $n_1 = n_0$. Conversely, the greater the ratio of the indices the stronger the scattering. Since R_θ is a function of the particle size, one can also in principle obtain the particle volume, or radius if the particles are spherical, by means of light-scattering measurements. Because of the strong dependence of I/I_0 on wavelength, the scattering is strongest in the blue region of the spectrum: $(\lambda\,\text{red}/\lambda\,\text{blue})^4 \simeq 7$. Thus, if a colloid dispersion is viewed in white light, the scattered light in the Tyndall cone has a blue appearance, while the transmitted light is richer in red light. This accounts for the blue of the sky and the redness of the setting sun – phenomena for which Rayleigh was originally seeking an explanation. Here the scattering centres are in fact molecules rather than colloid particles, although for a considerable time there were arguments as to the nature of the scattering units: ice particles, dust, and water droplets were all postulated as causing the effect.

The total scattered intensity (the *turbidity*, τ) is obtained by integrating the scattered intensity over the surface of a sphere to give

$$\tau = \frac{I_{\text{scatt}}}{I_0} = \frac{24\pi^3}{\lambda^4}\left(\frac{n^2 - 1}{n^2 + 2}\right) \cdot v^2 c \tag{7.9}$$

or, in terms of the scattering at 90°,

$$\tau = \frac{16\pi r^2}{3} \cdot \frac{I_{90}}{I_0} = \frac{16\pi}{3} R_{90}. \tag{7.10}$$

It follows from equation (7.9) that

$$\tau \propto (vc)v \tag{7.11}$$

so that, if the total volume, vc, of particles is constant, the turbidity increases with the particle size. This is what is observed when flocs increase in size during flocculation, and it provides a method of studying the kinetics of flocculation.

Before discussing the limitations of Rayleigh's theory and its extensions, mention may be made of one other application. In the above discussion it has been assumed that the scattering is caused by a material particle. However, any local fluctuation in density (and hence in refractive index) in an otherwise homogeneous phase can cause scattering. Thus even pure liquids scatter light feebly. In solutions fluctuations in local concentration also lead to light scattering. It turns out that these fluctuations are dependent on the molar mass of the solute and on the extent to which the solution departs from ideality. By carrying out measurements at a series of concentrations and at various scattering angles and extrapolating to zero concentration and scattering angle it is possible to obtain the molar mass of the solute. This technique is applicable to solutions of macromolecules.

The main limitations of the Rayleigh theory are that it applies only to particles much smaller than the wavelength of light, takes no account of particle shape, and applies only to dilute suspensions.

Theories to overcome these limitations involve complex mathematical treatments. The effects of increased particle size and deviations from spherical shape can, however, be described in reasonably simple terms. The first effect of increasing particle size is to elongate the scattering diagram along the direction of illumination (Figure 7.6), with the result that scattering at 45° is no longer equal to that at 135° but is greater. Thus the first sign of departure from the Rayleigh equations is that the ratio I_{45}/I_{135} is greater than unity. The origin of this effect is that all parts of a particle are not, at a given instant, subjected to the same electric field: light scattered from different regions within the particle is out of phase and therefore interferes at the detector, reducing the intensity of scattering.

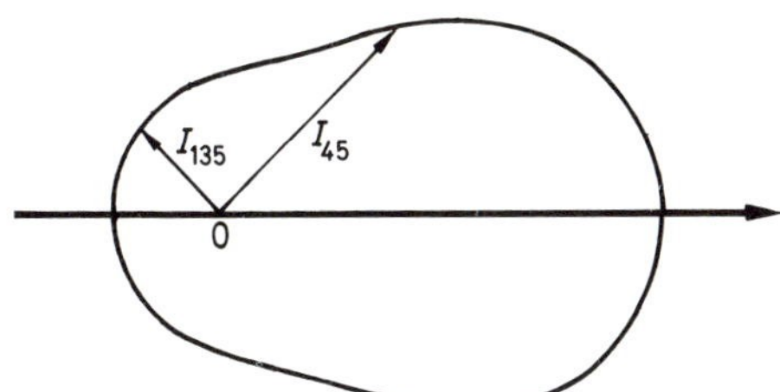

Figure 7.6 *Intensity of light scattered from an unpolarised beam by a large spherical particle at the origin, as a function of scattering angle θ (Rayleigh–Gans–Debye theory). The intensity of light scattered at 135° is less than that at 45°.*

The theory to take account of these effects was developed mainly by Debye and by Gans and applies when $(n_1 - n_0)R/\lambda \ll 1$; it is usually referred to as the Rayleigh–Gans–Debye (RGD) theory. The Rayleigh scattering (I_R) is modified by a correction factor $P(\theta)$:

$$I_{RGD} = I_R P(\theta), \tag{7.12}$$

where $P(\theta)$ can be expressed as a series expansion:

$$P(\theta) = 1 - \tfrac{1}{3}(QR_G)^2 + \dots \tag{7.13}$$

Here $Q = (4\pi/\lambda)\sin(\theta/2)$ and is called the *wave vector*, while R_G is the *radius of gyration* of a particle of arbitrary shape [for a homogeneous sphere $R_G = (\frac{3}{5})^{1/2}R$]. This shows that when $\theta = 0°$ $\sin(\theta/2) = 0$ and $P(\theta) = 1$, while when $\theta = 180°$ $\sin(\theta/2) = 1$ and $P(\theta) = 1 - \frac{1}{3}(4\pi R_G/\lambda)^2 + \ldots$; thus scattering in the backwards direction is decreased compared with Rayleigh scattering. To a first approximation, when $QR_G \ll 1$ and only the first term in the series is important,

$$\frac{I_{45}}{I_{135}} = \frac{1 - 0.049(4\pi R_G/\lambda)^2}{1 - 0.285(4\pi R_G/\lambda)^2}, \tag{7.14}$$

thus enabling R_G to be calculated from measurements at 45° and 135°.

When the particles are much larger, $(n_1 - n_0)R/\lambda \simeq 1$, the theory becomes very complex (*Mie theory*), especially when account is taken of the absorption of light by the particle. The main consequence of Mie's theory is that the scattering diagram now shows a complex dependence on θ; an example is shown in Figure 7.7. However, by computation of such patterns for a range of size and shape parameters it is now possible to match experimental data to theoretical scattering profiles and to derive information on particle size and shape. For example, the sizing of monodisperse spherical particles up to about 2 μm is possible by this technique.

An outstanding limitation still remains: in more concentrated dispersions light scattered by one particle may undergo further scattering encounters with other particles before leaving the sample. These multiple scattering effects soon become serious, and no adequate theory is yet available to describe them. In practice they are mitigated by working with dilute suspensions and small sample volumes.

Dynamic Light Scattering

In conventional light scattering the light scattered from different particles is of random phase so that destructive interference does not occur in a systematic way. If, however, a laser is used, the light is coherent (*i.e.* the wave motion is in phase across a plane of propagating light) and interference patterns are produced by an array of particles. If the particles are stationary, then a pattern of bright spots is seen: Figure 7.8 shows diffraction patterns

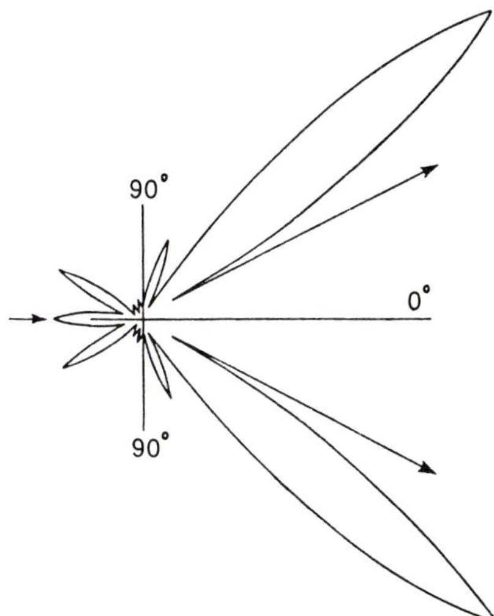

Figure 7.7 *Complex scattering pattern for a very large spherical particle (Mie theory).*

obtained by shining a laser beam through thin films of polymer particles deposited on a glass plate. Figure 7.8(a) is analogous to the Laue *X*-ray diffraction pattern produced by a crystal.

In a colloidal dispersion particles are in continuous motion and the pattern of spots changes continuously, leading to a mobile speckle pattern. If attention is directed to a specific area of view, the light appears to twinkle as bright spots appear and disappear from that area. The character of the twinkling is clearly related to the motion of the particles, and dynamic light scattering exploits this fact. The nature of the twinkling is sensed by a 'photon correlator', which enables the correlation in time between the fluctuations in light intensity to be analysed; for this reason the technique is frequently described as *photon correlation spectroscopy* (PCS). The result is a correlation function which represents the relation between the average intensity at a time $(t + \tau)$ and that at time t. When τ is large, $I(t)$ and $I(t + \tau)$ are independent of one another, while at very short delay times they are closely related. This function in effect is a measure of the probability of a particle moving a given distance in a time τ. In the case of equal-size spherical particles the correlation function is a simple exponential function:

$$g(\tau) = \exp(-\tau/\tau_c), \tag{7.15}$$

where the parameter τ_c is related to the diffusion coefficient, D, of the particles:

$$\tau_c = 1/DQ^2. \tag{7.16}$$

By plotting $\ln g(\tau)$ against τ, all of which is done by computer, τ_c is found. Assuming the Stokes–Einstein equation for the diffusion coefficient [equation (6.10)], the hydrodynamic radius, R, of the particles is given by

$$R = (kT/6\pi\eta)Q^2\tau_c. \tag{7.17}$$

In practice one measures τ_c at various scattering angles (various values of Q) and plots $1/\tau_c$ against Q^2 to obtain D and hence R.

PCS is now widely used for the determination of the size of particles in the colloid range. In general the results are in reasonably good agreement with those found by electron microscopy. Such discrepancies that do occur may be attributed either to a difference between the hydrodynamic radius (which may include a sheath of solvent or an adsorbed layer on the surface of the particles) and the real radius or to changes in the particle size arising from the preparation (drying) of the sample for electron microscopy or arising from bombardment by the electron beam. The method is, however, much less easily applied to polydisperse systems, since the correlation function then has a much more complex form and has to take account of the movement of particles of different sizes.

Another application of laser light scattering is to the study of electrophoresis. This makes use of the Doppler effect. If light is scattered by a moving particle, the wavelength of the scattered light is changed. Thus, by comparing the wavelengths of the scattered and incident light, the drift velocity of the charged particles under the influence of the electric field can be determined. The method has several important advantages in speed and accuracy over conventional observational methods of measuring electrophoretic mobility (see Chapter 6).

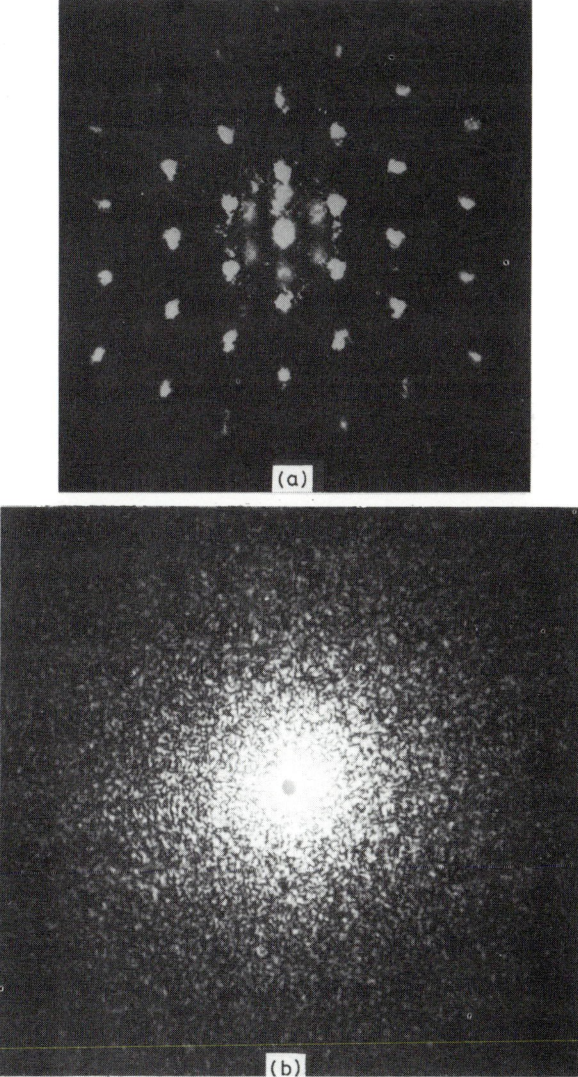

Figure 7.8 *Interference patterns produced by passing a laser beam through a thin layer of polymer latex particles deposited on a microscope slide:* (a) *highly ordered layer showing Laue pattern,* (b) *disordered layer showing a speckle pattern which twinkles if the particles are mobile. The light intensity from a given area then fluctuates because of the Brownian motion of the particles.*

[(a) Reproduced with permission from Professor R. H. Ottewill and (b) photographed by D. Jones, School of Chemistry, University of Bristol]

NEUTRON SCATTERING

Because of the dual wave–particle character of subatomic particles, a beam of neutrons behaves in a wave-like fashion in that, when scattered by matter, the scattering patterns that are produced are analogous to those obtained by light or *X*-rays. However, the effective wavelength of neutrons is very small – in the range 1—10 Å depending on their velocity – so that when applied to scattering by colloid-size particles one has $R/\lambda \gg 1$. The equation representing the scattering is therefore of the same form as equations (7.3) and (7.12):

$$(I/I_0) = A(\rho_p - \rho_m)^2 v^2 P(Q). \tag{7.18}$$

Here the term $(\rho_p - \rho_m)^2$ replaces the refractive index term in equation (7.3): ρ_p and ρ_m are, respectively, the 'neutron scattering-length densities' of the particle and the medium. A is a constant incorporating the other factors in equation (7.3) and is a characteristic of the apparatus, *i.e.* the wavelength of the neutrons, the distance from target to detector. In this case the simple equation for $P(Q)$ given in equation (7.13) no longer applies. For spheres it now has the form

$$P(Q) = [(3 \sin QR - QR \cos QR)/(QR)^3]^2 \tag{7.19}$$

and exhibits an oscillating character, illustrated in Figure 7.9. The separation of the peaks is a measure of the particle size. In practice the peaks are much less well defined than those shown in the figure; nevertheless one can obtain satisfactory values for the mean particle size.

It can be seen from Figure 7.9 that the scattering angles used in neutron scattering are very small – a matter of a few degrees – so the technique is known as *small-angle neutron scattering* (SANS).

This technique, although not superior to ordinary light scattering for the determination of particle size, has a number of important advantages over light scattering when applied to more complex problems. Two examples may be mentioned briefly.

The first arises from the fact that the neutron scattering-length densities of different atoms vary widely: in particular those for hydrogen and deuterium differ considerably so that by mixing H_2O and D_2O in varying proportions media of widely different values of ρ can be prepared. Equation (7.18) shows that if $\rho_p = \rho_m$ there is no scattering, *i.e.* neutrons do not 'see' the particles. This principle can be exploited in the study of composite particles, such as particles covered with an adsorbed layer. One may either match the medium to the adsorbed layer, to obtain the size of the core, or match the core when the thickness of the adsorbed layer can be estimated. In this way a method may be developed for the measurement of adsorption from solution by colloid particles.

A second important application of neutron scattering is to concentrated dispersions which cannot be studied by light scattering because of their opacity. The method has close analogies with the use of low-angle *X*-ray scattering to study the structure of liquids, and its main application so far has been to investigate the radial distribution function for colloid particles in concentrated dispersions.

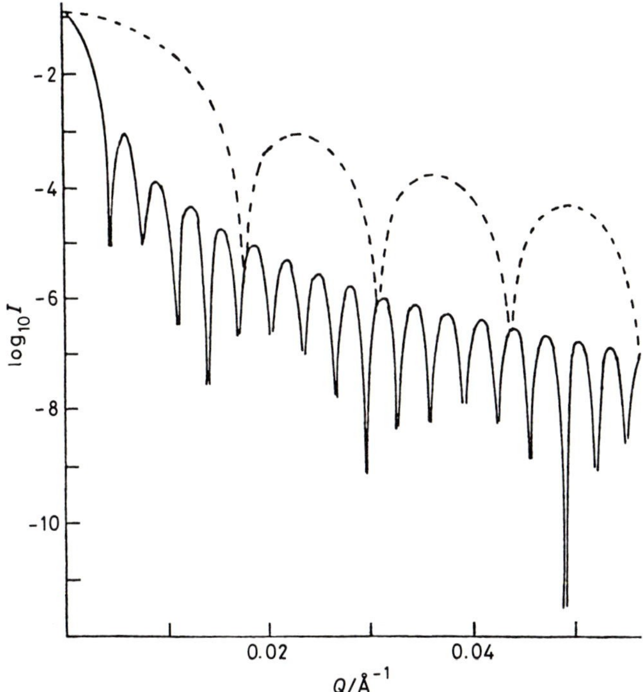

Figure 7.9 *Log I as a function of Q for spherical particles according to equation (7.19). Full curve, particle radius 100 nm; dashed curve, particle radius 25 nm. For neutrons of wavelength 5 Å Q = 0.05 Å, corresponding to a scattering angle of 2.6°.*

(Reproduced from 'Colloidal Dispersions', ed. J. W. Goodwin, Special Publication No. 43, The Royal Society of Chemistry, London, 1982)

Chapter 8

Some Important Properties of Colloids III Rheology

INTRODUCTION

Rheology is concerned with the flow and/or deformation of matter under the influence of externally imposed mechanical forces. Two limiting types of behaviour are possible. The deformation may reverse spontaneously (relax) when the external force is removed: this is called *elastic behaviour* and is exhibited by rigid solids. The energy used in causing the deformation is stored, and then recovered when the solid relaxes. At the other extreme, matter flows and the flow ceases (but is not reversed) when the force is removed: this is called *viscous behaviour* and is characteristic of simple liquids. The energy needed to maintain the flow is dissipated as heat. Between the two extremes are systems whose response to an applied force depends on the time-scale involved. Thus pitch behaves as an elastic solid if struck but flows if left for years on a slope. Similarly, a ball of 'Funny Putty', a form of silicone rubber, bounces when dropped on a hard surface, when the contact time is a few milliseconds, but flows if deformed slowly on a time-scale of seconds or minutes. Systems of this kind are said to be *visco-elastic*. The precise nature of the observable phenomena depends on the ratio of the time it takes for the system to relax to the time taken to make an observation. This ratio is called the *Deborah number** (*De*):

* The term Deborah number was coined by M. Reiner (see *Physics Today*, 1969, **17**, 62), who recalled the song of the prophetess Deborah (*Judges*, Chapter 5, verse 5) 'The mountains melted ['flowed' in the Hebrew] from before the Lord', suggesting that given infinite time (God's time) even mountains will flow.

$$De = \text{time for relaxation/time of observation.} \tag{8.1}$$

When De is very large, the system behaves like a solid and has elastic properties: when De is very small, it flows like a liquid. The importance of rheology in studying colloidal dispersions arises from the fact that De is often around unity so that a wide range of visco-elastic phenomena are observed.

Flow or deformation involves the relative motion of adjacent elements of the material. As a consequence such processes are sensitive to interatomic or intermolecular forces. In the case of liquids containing dispersed particles, interparticle forces are also important. Because the rheological properties of colloidal suspensions exhibit such a rich variety of phenomena, rheological studies not only provide information on medium–particle and particle–particle interactions but also are of immense technological importance.

It is not possible in this chapter to do more than give a brief review of the subject and to indicate in rather general terms the way in which rheological measurements can help us to understand the behaviour of colloids. Some of the technological applications of rheological properties will be dealt with in Chapter 14.

VISCOSITY

We shall limit discussion to the simplest case of *shear viscosity*, illustrated in Figure 8.1(a). Fluid is contained between two plates parallel to the xy-plane and a distance h apart along the z-axis. The plates are maintained in relative motion at a constant velocity V in the x-direction by a shearing force F applied to the upper plate and a force $-F$ to the lower plate.

Assuming that there is no slip at the solid/liquid interfaces, the velocity of movement, v, of an element of fluid relative to the lower plate increases linearly from zero at $z = 0$ to V at $z = h$, as shown in Figure 8.1(b). The ratio V/h is called the *shear rate*, or the *rate of shear strain*, and is denoted* by D. The force needed to maintain the steady motion is proportional to the area, A, of the plates (ignoring edge effects), and the ratio F/A is called the *shear*

* The notation and nomenclature employed are those recommended in the IUPAC Manual of Symbols and Terminology for Physico-chemical Quantities and Units, Appendix II, Part 1.13 (Selected Definitions, Terminology and Symbols for Rheological Properties), *Pure and Applied Chemistry*, 1979, **51**, 1213–8. Alternatives to D and σ are respectively $\dot{\gamma}$ and τ.

stress and is denoted by σ.

The ratio σ/D is called the *viscosity* (η):

$$\eta = \sigma/D, \tag{8.2}$$

while its reciprocal is the *fluidity*.

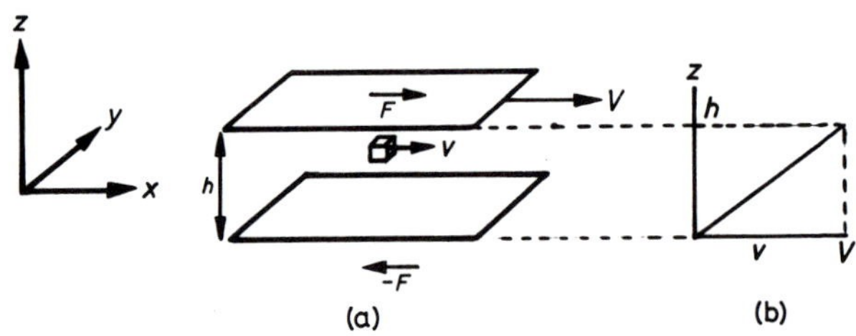

Figure 8.1 (a) *Representation of a simple shear field between two parallel plates.* (b) *Variation of the local velocity with distance in the space between the plates.*

In practice the idealised situation of parallel plates is approximated by using two co-axial cylinders whose radii are r_1 and r_2, where $(r_2 - r_1) \ll r_1$ [Figure 8.2(a)]. One cylinder is rotated at constant speed while the torque needed to prevent the other cylinder rotating is measured as a function of the speed of rotation. An alternative geometrical arrangement in which the shear rate is constant consists of a cone and plate [Figure 8.2(b)]. As one moves away from the axis of rotation both the velocity and the plate separation increase in proportion to the distance from the axis, and the ratio v/h remains constant. Again the torque required to hold the plate stationary when the cone rotates is measured. By far the simplest apparatus is the capillary viscometer [Figure 8.2(c)], in which one measures the rate of flow of liquid under the influence of a more or less constant hydrostatic head. However, the shear rate varies across the tube so that the method is only useful when the viscosity is independent of the shear rate. A further, less precise, method is that in which the terminal velocity of a solid sphere falling through the liquid is measured [Figure 8.2(d)]. The rate of rise of a small bubble may also be used in the same way.

NEWTONIAN AND NON-NEWTONIAN SYSTEMS

In the simplest type of behaviour, exhibited by *Newtonian systems*,

η is a constant independent of D, as shown in Figure 8.3(a). Simple molecular liquids and dilute colloidal systems usually respond in this way.

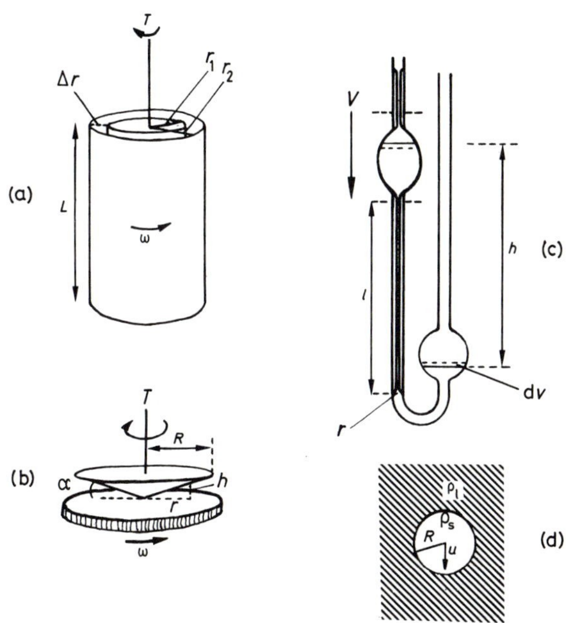

Figure 8.2 *Basic methods of measuring viscosity:* (a) *concentric cylinders (Couette)* $[T = (2\pi L r^3/\Delta r)\omega\eta]$, (b) *cone and plate* $[T = \frac{2}{3}\pi R^3(\omega/\alpha)\eta]$, (c) *capillary viscometer (Ostwald)* $[\eta = (\pi r^4 \rho g h/8l)/(dv/dt)]$, (d) *falling sphere (Stokes)* $[\eta = \frac{2}{9} g R^2(\rho_s - \rho_l)/u]$.

In *non-Newtonian systems* the viscosity is shear dependent: we then talk of the *apparent viscosity*, η_{app}, or *shear-dependent viscosity*, defined by equation (8.2). Some typical types of behaviour are illustrated in Figure 8.3. The apparent viscosity, given by the slope of the dashed lines, may increase with D [Figure 8.3(b)], when the fluid is said to exhibit *shear thickening* (sometimes called *dilatency*), or, in the case of *shear thinning*, it may decrease [Figure 8.3(c)]. A more extreme case of shear thinning, called *pseudo-plastic* behaviour, is shown in Figure 8.3(d). In *plastic systems* [Figure 8.3(e)] there is no initial response, on the time-scale of observation, to the imposition of stress, but flow begins at a *limiting yield stress* (σ_0). It is essential to specify the time-scale involved. Thus σ_0 for pitch is high if the force is applied quickly, but a small force applied over years causes flow. Many colloidal dispersions show *Bingham flow*, in which, at higher shear rates, the relation between σ and D becomes linear, the slope being the *differential viscosity*, η_Δ (sometimes called the *plastic viscosity*, η_{pl}). In this region

$$\sigma - \sigma_{\mathrm{B}} = \eta_\Delta D, \tag{8.3}$$

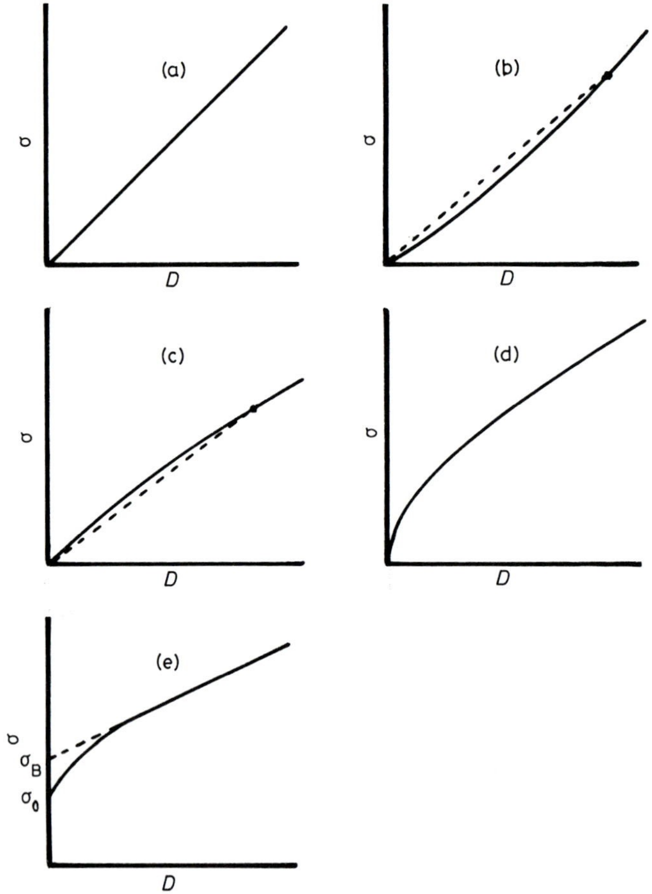

Figure 8.3 *Basic types of rheological behaviour:* (a) *Newtonian,* (b)—(e) *non-Newtonian [*(b) *shear thickening,* (c) *shear thinning,* (d) *pseudoplastic,* (e) *plastic (Bingham flow), in which σ_0 is the yield stress and σ_B is the Bingham yield stress].*

the intercept of the line extrapolated from the linear region with the σ-axis being called the *Bingham yield stress* (σ_B).

As emphasised above, the rheological properties of colloid systems are often time dependent. If after a long rest the application of a finite shear leads to a decrease in viscosity, *work softening* is said to occur; if on standing the initial higher viscosity is regained, the system is said to be *thixotropic*. The time taken for this to occur is the *time of thixotropic recovery*. In other cases opposite effects may be observed. *Work hardening* is said to take place if the viscosity increases with the length of time for which the stress is applied.

RHEOLOGY OF SUSPENSIONS OF COLLOIDAL PARTICLES

The Newtonian viscosity of a liquid is modified, and may become non-Newtonian, if it contains colloidal particles. This results from the complex interplay of interactions including hydrodynamic interactions between the liquid and solid particles, attractive or repulsive forces between the particles, and, in concentrated systems, direct particle–particle contact.

Hydrodynamic interactions between the liquid and solid particles suspended in it result essentially from the finite volume occupied by the particles: they are excluded volume effects leading to additional viscous dissipation in the liquid and are significant even at low concentrations.

Spherical Particles

According to a simple theory due to Einstein (1906) for spherical particles at very low concentrations, the *relative viscosity*, η/η_s, where η_s is the viscosity of the pure liquid medium, is related to the volume fraction, ϕ, of particles by the simple equation

$$\eta/\eta_s = 1 + 2.5\phi. \tag{8.4}$$

This may also be expressed in terms of the *relative viscosity increment*, η_i (the term *specific viscosity* is now discouraged by IUPAC):

$$\eta_i = (\eta - \eta_s)/\eta_s = 2.5\phi \tag{8.5}$$

or

$$[\eta] = \eta_i/\phi = 2.5, \tag{8.6}$$

where $[\eta]$ is often called the *intrinsic viscosity*.* In general, the intrinsic viscosity is the limit of η_i/ϕ as $\phi \rightarrow 0$.

Einstein's equation is strictly valid at infinite dilution where only single particle/medium interactions are important. It is found in practice to be followed with reasonable accuracy up to volume fractions of about 0.01. The origin of this effect can be understood qualitatively in the following way [Figure 8.4(a)]. The centre of mass of a particle moves with a velocity v, equal to the flow velocity of the medium at this level. In the absence of the particle the medium at the level corresponding to the upper boundary of the particle would move at a velocity $v + R(dv/dz) = v + RD$, where R is the radius of the particle.

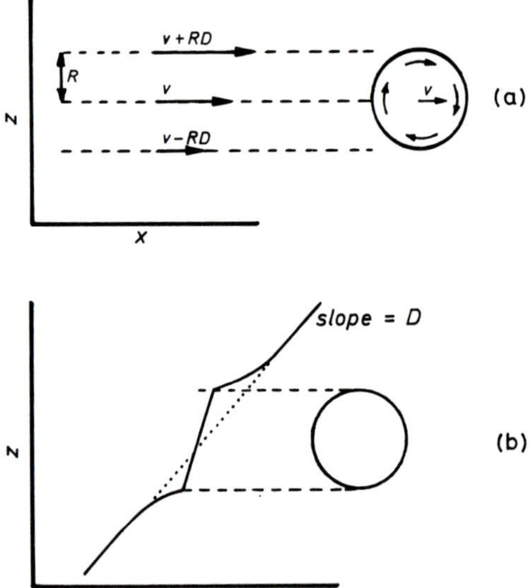

Figure 8.4 (a) *Rotational motion of a sphere induced by a shear field: this motion is resisted by viscous effects over the surface of the sphere.* (b) *Resultant distortion of the flow field of the liquid: dotted line – flow field in the absence of the sphere; full line – effect of the presence of the sphere, which slows down the liquid around the upper hemisphere and accelerates it in the lower.*

* This term is also used for η_i/ρ, where ρ is the mass fraction of particles.

At the lower boundary its velocity would be $v - RD$. When a particle is present, this will no longer be the case. If there is no slip at the liquid/solid surface, the forces acting on the upper and lower hemispheres of the particle will cause it to rotate. However, this rotation will be resisted by viscous forces over the whole surface of the particle; for example, there will be shears down and up, respectively, on the leading and trailing surfaces. Consequently the velocity of the liquid in contact with the top surface of the sphere will be lowered and that at the bottom increased. The velocity profile of the fluid round the perimeter of the particle will be distorted as indicated schematically in Figure 8.4(b). If the overall flow is to be kept constant, the stress must increase, *i.e.* the viscosity is increased by the presence of the particle. In a dispersion the total effect will be proportional to the concentration of particles and to their volume, leading to the relationship in equation (8.5).

At higher concentrations one has to consider the effects of pairs of particles in close proximity. For example, if two particles are at different levels in the flow field, that at the higher level will overtake and pass that at the lower level [Figure 8.5(a)]. If their two paths are close together, and separated by less than $2R$, the trajectories of the particles will be distorted as shown in Figure 8.5(b). This change of trajectory involves an expenditure of energy. Since the probability of close approach of two particles is proportional to the square of the volume fraction, a correction term proportional to ϕ^2 must be added to equation (8.5). The exact theoretical treatment of this effect is complicated since account has to be taken both of the hydrodynamic forces when two particles pass one another on near-miss trajectories and of the influence of the Brownian motion of the particles. Recent theoretical calculations have shown that the proportionality factor is 6.2. On raising the particle concentration further, the influence of three or more neighbouring particles must be included. It is then reasonable to expect equation (8.5) to represent only the first term in a series expansion in ϕ:

$$\eta_i = 2.5\phi + 6.2\phi^2 + k_3\phi^3 + \ldots \quad (8.7)$$

Only very approximate values of k_3 have been derived, and little is known about the higher terms. According to equation (8.7) the ratio η_i/ϕ will have the form shown in Figure 8.6.

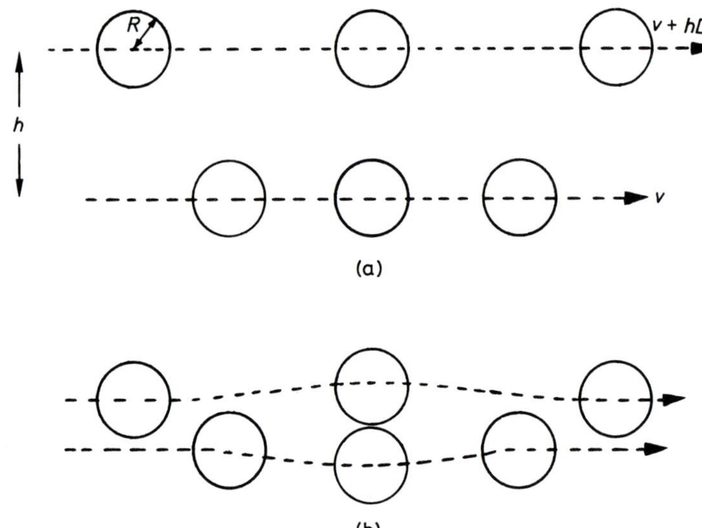

Figure 8.5 *Relative motion of particles at different levels in the flow field. Positions are shown at successive time intervals:* (a) *when the trajectories are separated by substantially more than the collision diameter ($h \gg 2R$),* (b) *when the trajectories are separated by less than the collision diameter ($h < 2R$), showing distortion of the trajectories. (Brownian motion is neglected in this simplified representation.)*

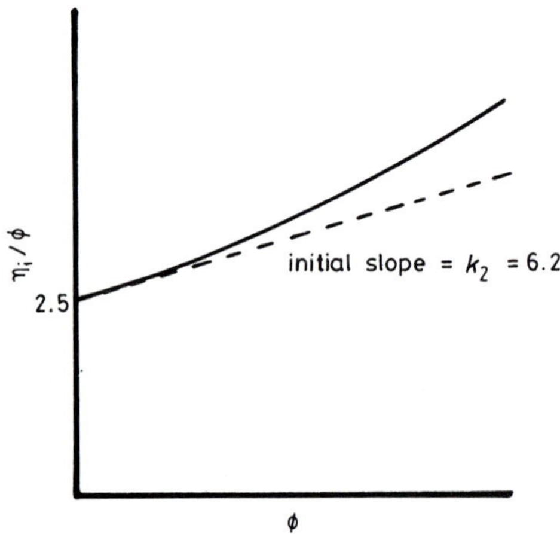

Figure 8.6 *Variation of η_i/ϕ with ϕ for a system in accordance with equation (8.7).*

So far we have taken no account of interparticle forces. In the absence of such forces two particles at different levels at a separation greater than $2R$ will pass one another on essentially parallel paths as already shown in Figure 8.5(a). However, if the particles attract one another, their trajectories will be modified in the way illustrated in Figure 8.7(a). Again there will be an additional contribution to the drag and an increase in viscosity. The effect increases with the increasing magnitude of the interparticle forces. If they are strong enough, the particles may form a doublet – corresponding, for example, to secondary minimum flocculation – which will be induced to undergo rotatory motion [Figure 8.7(b)]. This represents yet another mode of energy dissipation which contributes to the viscosity. Since pairs of particles are involved, interparticle forces leading to reversible doublet formation should not affect the intrinsic viscosity.

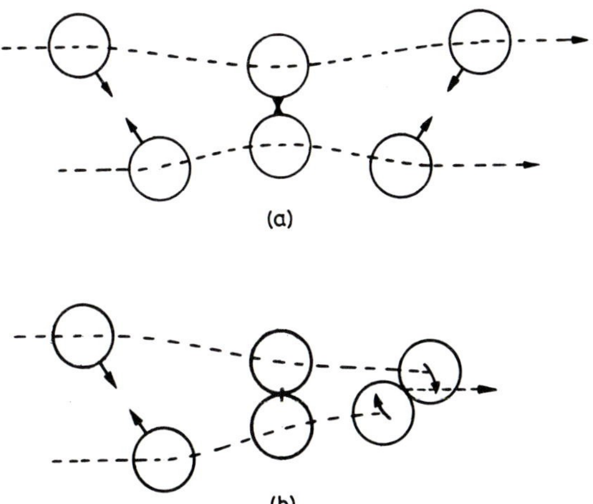

Figure 8.7 *Trajectories of attracting colloidal particles in a flow field. As in Figure 8.5, the positions of the particles are shown at successive intervals of time.* (a) *Moderate attractive forces leading to a distortion of the trajectories,* (b) *strong attractive forces leading to the formation of a doublet which then rotates in the field. (Brownian motion is neglected in this simplified representation.)*

Increase in concentration, or of interparticle forces, will lead to larger aggregates which will enhance the viscosity even further. Systems under these conditions are usually non-Newtonian since these effects are sensitive to shear rate. Increase in the shear rate will tend to dissociate the flocs, which will decrease in both concentration and size, and the system will exhibit shear thinning. Yet another effect arises at very high shear rates. In Chapter 9 mention will be made of *orthokinetic flocculation*, in which hydrodynamic forces can bring particles together with sufficient energy to overcome the primary maximum and lead to primary minimum coagulation. If this occurs, then shear thinning at lower shear rates may be followed at very high shear rates by shear thickening caused by the presence of flocculated particles.

Increases in both particle concentration and magnitude of interparticle forces may lead to an increase in the size of flocs to such an extent that they form a particle network (*gel*, see Chapter 13) which resists flow until the yield stress is reached. The system becomes first pseudo-plastic and then plastic.

Other phenomena may also be observed in concentrated dispersions. Thus, as the concentration increases, direct particle–particle contacts increase in importance, and as the limiting packing fraction is approached ($\phi = 0.625$ for the random close-packing of spheres) the suspension ceases to flow and the viscosity becomes infinite. Shortly before this limit is reached the suspension (especially if the particles are monodisperse spheres) often exhibits dilatency. When the particles are nearly close-packed, flow can occur only by the particles rolling past one another (Figure 8.8), and this results in an increase in volume: the shear thickening is accompanied by dilatence of the system. If the amount of liquid present is insufficient to fill the extra void volume produced, the surface may become dry. This is the familiar phenomenon observed when one steps on wet sand.

Figure 8.8 *Schematic representation of the dilation caused by the relative motion of particles in a nearly close-packed system; note the increase in the void volume (shaded).*

Systems of this kind often show additional interesting and important time effects. Thus, if the stress is applied rapidly, the particle structure does not have time to relax and the material appears to be almost solid. On the other hand, slow application of stress leads to steady flow, or *creep*. The suspension can be stirred slowly, but it develops a very high viscosity if disturbed suddenly. The non-drip properties of many modern paints depend on such thixotropic phenomena.

In many colloid systems particles are covered with adsorbed layers (see Chapter 5). These too influence the viscosity since the effective radius, and hence the effective volume fraction, is greater than that of the core particles. In attempting to fit experimental data on dispersions of spherical particles to theoretical equations the effective volume fractions must be employed. If measurements are made on very dilute suspensions and at low shear rates, equation (8.7) (retaining only the first two terms) may be used to calculate the effective volume fraction and hence the particle size, and the thickness of the adsorbed layer if the size of the core particles is known. This is not, however, a very precise method and generally other methods of finding the adsorbed layer thickness are to be preferred.

Non-Spherical Particles

If the particles are strongly anisotropic then, as illustrated in Figure 8.9(a), their rotational motion in the shear field is greatly enhanced. Additional energy is dissipated in maintaining this motion and the viscosity is increased. Since only single particles are concerned in this process, it will increase the intrinsic viscosity above the value of 2.5 for spherical particles. For example, it is calculated that for rods having an axial ratio of 15 the intrinsic viscosity rises to 4.0.

An important feature of the rotation of a rod-like particle is that the turning moment depends on its orientation relative to the direction of the shear field. The moment, and hence the angular velocity, is greatest when the particle is at 90° to the shear field and least when it is parallel to it [Figure 8.9(a)].* As the axial ratio increases, this variation of angular velocity with orientation increases dramatically until, although particles continue to rotate, they spend most of their time in a near-parallel orientation

* In the parallel orientation the turning moment is zero, so one might expect rotation to cease. However, Brownian motion will ensure that from time to time the particle will rotate slightly and then be subjected to forces causing it to rotate further.

[Figure 8.9(b)]. Their rotation consists of occasional 'flicks' through 180°. Their influence on the viscosity is decreased and the flowing liquid is no longer optically isotropic but becomes birefringent; the phenomenon is known as *flow birefringence* and its appearance is clear evidence that the particles are strongly anisotropic [Figure 8.9(c)].

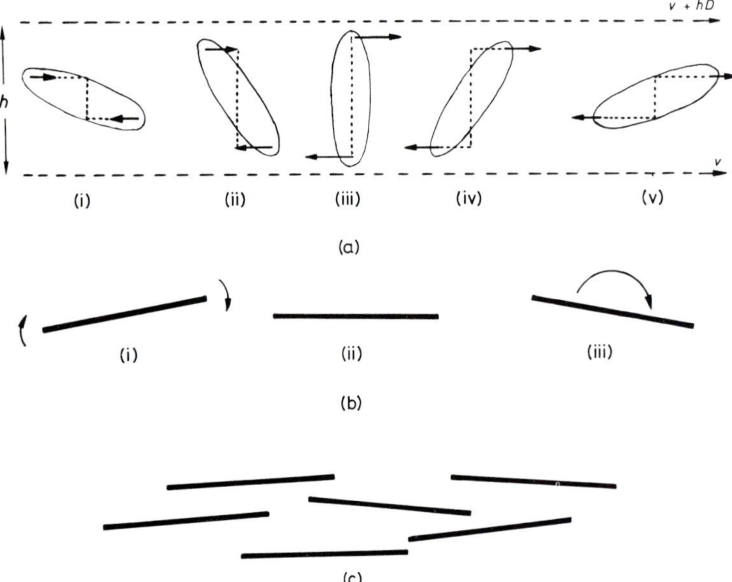

Figure 8.9 *Behaviour of anisotropic particles in a shear field.* (a) *Rotation induced by a shear field: the turning moment, and hence the angular velocity, is greatest when the major axis is at 90° to the field* (iii). (b) *If the axial ratio is very large, the predominant orientation is parallel to the field* (ii). *If Brownian motion causes the particle to rotate slightly, the flow may either, as in* (i), *restore the parallel orientation or, as in* (iii), *lead to a rapid 'flick' through 180°.* (c) *Optical anisotropy results when a large number of long rod-like particles are oriented in the shear field.*

In view of the complexity of the possible rheological phenomena which colloidal systems can exhibit it is not surprising that, even for dispersions of spherical particles, equation (8.7) holds only up to a volume fraction of about 0.3 and that a large number of empirical equations have been suggested to represent real systems at high concentrations. One semi-empirical equation that is often useful and gives a reasonable representation of shear thinning is the following:

$$\ln(\eta/\eta_s) = -[\eta]\phi_p(1 - \phi/\phi_p), \tag{8.8}$$

where ϕ_p is the maximum packing fraction, at which the viscosity becomes infinite. For equal-size spheres one might expect ϕ_p to be about 0.6 and $[\eta] = 2.5$. In practice, for real systems that follow this equation both parameters differ from these values, and they are often shear dependent.

So far, little progress has been made in understanding the rheological properties of polydisperse systems, especially at high concentrations.

ELECTROVISCOUS EFFECTS

As outlined earlier (Chapter 3), electrically charged colloidal particles are surrounded by an electrical double layer. This leads to important rheological effects. The *primary electroviscous effect* is a consequence of the fact that the electrical double layer [Figure 8.10(a)] is deformed from its spherical shape by the shear field. Construction of the double layer ahead of the particle and its disintegration behind the particle take a finite time [Figure 8.10(b)]. This causes an increase in the intrinsic viscosity, which for low zeta-potentials ($\zeta < 25$ mV) is proportional to the square of this potential:

$$[\eta] = 2.5 + f(c)\zeta^2, \tag{8.9}$$

where $f(c)$ is a function of both the ionic concentration c in the bulk and the particle size. Furthermore, the repulsion between the double layers increases the effective collision diameters of the particles and hence their effective volume, whilst also leading to distortion of the particle trajectories [Figure 8.10(c)]. This gives rise to the *secondary electroviscous effect*, which again results in an increase in the viscosity. As mentioned above, the presence of adsorbed layers on a particle also increases its effective diameter. If these adsorbed layers are polyelectrolyte molecules, then their effective thickness depends on both salt concentration and pH [Figure 8.10(d)]: this is the origin of the *tertiary electroviscous effect*. Changes in the solution composition that lead to a more compact adsorbed layer decrease the viscosity, and *vice versa*.

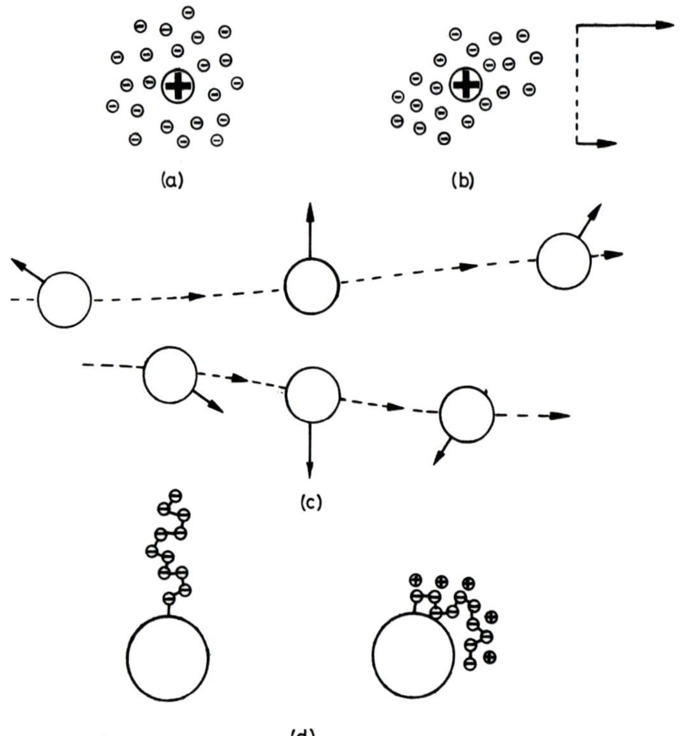

Figure 8.10 *Origin of electroviscous effects:* (a) *electrical double layer round a particle at rest,* (b) *distortion of the electrical double layer in a shear field, leading to the primary electroviscous effect,* (c) *trajectories of repelling particles caused by double-layer repulsion, leading to the secondary electroviscous effect,* (d) *effect of ionic strength (or pH) on the extension of a charged adsorbed polyelectrolyte, causing a change of the effective diameter of the particle, and the tertiary electroviscous effect.*

RHEOLOGICAL SPECTROSCOPY

A rapidly expanding field of study is that concerned with the time response of colloidal systems to a mechanical disturbance. The importance of this was hinted at at the beginning of the chapter (page 110), and recent major developments in instrumentation have opened up new areas of research.

Referring to Figure 8.1, we suppose that the plates are initially at rest and that at a time t_0 the upper plate is subjected to a rapid, short-lived displacement. The resultant stress on the lower plate will not be experienced immediately, but only after a certain delay time (Δt). By measuring the delay time as a function of the separation of the plates, the velocity of propagation of the shear wave can be measured. This is the principle of the *shear rheometer*. The information obtained from such experiments gives significant new information concerning the rheological properties of the intervening material.

A technique of longer standing is that in which one component of a rheometer [Figure 8.2(a) or (b)] is caused to undergo a sinusoidal oscillation and the response of the other is monitored (Figure 8.11). The recorded stress curve at the inner cylinder, or the cone, lags behind the strain curve imposed on the outer cylinder, or the lower plate. The variation of the relaxation time with frequency provides information on the processes responsible for the dissipation of energy which occurs as the individual elements of material between the plates respond to the continuously varying stress to which they are subjected. The principles involved in this technique are essentially similar to those governing the response of dielectric materials to a fluctuating

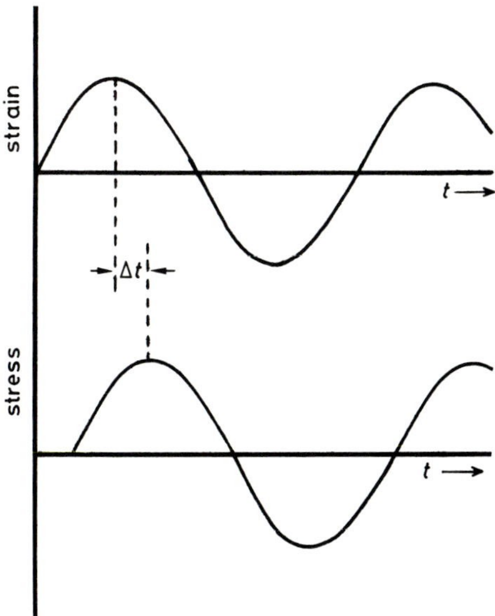

Figure 8.11 *Time lag between the response of stress to a sinusoidal imposed strain. The dependence of* Δt *on the frequency provides information on the kinetics of interparticle motions.*

electric field, and the results are analysed in an analogous fashion to identify the rheological storage and loss factors. These in turn can be related to the nature of the interparticle forces in a colloidal suspension.

Further discussion of rheological spectrometry would lead to a somewhat more involved treatment of response theory, which is beyond the scope of this book.

CONCLUSION

The great variety of rheological phenomena exhibited by colloidal dispersions, only a few of which have been dealt with above, not only provide a major theoretical challenge but also present many opportunities for exploitation in industrial applications, some of which will be mentioned in Chapter 14.

Chapter 9

How are Colloidal Dispersions Destroyed? I Aggregation Processes

INTRODUCTION

Flocculation involves the aggregation of particles without the destruction of their individuality and differs fundamentally from the processes of sintering and particle growth, to be discussed in the next chapter. Since flocculation is the first step in the destruction of a very wide range of colloids, it is important to understand the underlying mechanisms involved. Major research efforts have and are being made to elucidate the details of this process.

As explained in Chapter 2, flocculation occurs spontaneously if it is accompanied by a decrease in the total free energy of the system. This free energy is made up of several contributions. We have seen earlier that one contribution arises from the potential energy of interaction between colloidal particles which may be regarded as lowering the surface tension between the interacting surfaces. We denote this by $\Delta G(\text{int})$. In addition, the coming together of several particles to form an aggregate results in the loss of the independent translational motion of the individual particles and its incorporation in the translational motion of the aggregate (Figure 9.1). This reduction in the number of degrees of freedom implies a decrease in the entropy of the system, $\Delta S(\text{agg})$, and hence an increase in the free energy: $\Delta G(\text{agg}) = -T[\Delta S(\text{agg})]$.

In many examples of rapid coagulation it is assumed that the interaction free energy dominates the process so that one can ignore the contribution from the entropy term. The flocculation can then be treated solely in terms of the interparticle free-energy diagrams such as those discussed in Chapter 2. In these cases flocculation results from the lowering or elimination of the free-energy barrier (Figure 2.8), which allows the system to pass over into a state of lower interaction free energy corresponding to the primary minimum, whose depth is usually so great that entropy effects can be neglected.

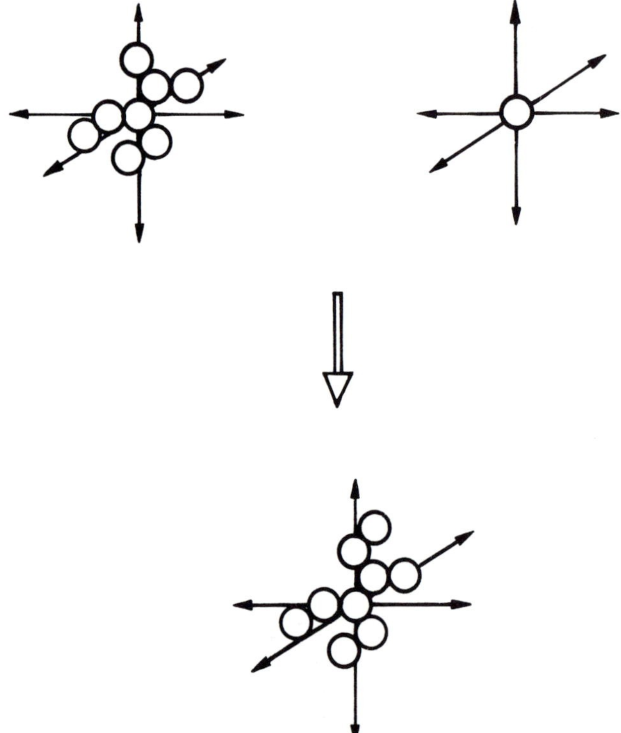

Figure 9.1 *Loss of degrees of translational freedom on addition of an extra particle to a floc.*

On the other hand, if the flocculation is caused by the presence of a secondary minimum, then entropy effects can be of major importance and lead to a variety of observed phenomena.

In this chapter we shall discuss first the coagulation of electrostatically stabilised aqueous dispersions in which entropy effects can be ignored and then systems involving secondary minima. Two other phenomena are then discussed briefly, namely *bridging flocculation* and *depletion flocculation*. This leads to a discussion of the kinetics of flocculation. Of increasing practical importance are the conditions under which mixtures of colloidal particles of different kinds flocculate – so-called *heteroflocculation*. The final section deals briefly with the important topic of the structure of flocs and sediments.

FLOCCULATION AND COAGULATION OF ELECTROSTATICALLY STABILISED DISPERSIONS

The study of the coagulation of aqueous dispersions of sparingly soluble or insoluble solids (*lyophobic colloids*) has been one of the main streams of colloid research for over a century. This arises both from its practical importance and from the fact that it was the first colloidal phenomenon for which empirical rules governing its onset were established. Consequently it was also the first for which attempts were made to develop a theoretical interpretation.

The coagulation of lyophobic colloids by the addition of electrolyte was observed by Faraday (1856) and was certainly known to others a good deal earlier. The converse situation in which precipitates of metal salts in analytical procedures tended to pass through a filter paper if the electrolyte concentration was too low (*peptisation*) was also familiar. The studies of Schultze (1882) and Hardy (1900) led to the general conclusion that flocculation was controlled mainly by the nature of the ion of the added electrolyte carrying a charge of opposite sign to the surface charge on the colloid particle and that under given conditions the effectiveness was strongly dependent on the valency of this *counter-ion*. The nature and valency of the ion of the same sign as the particle (*co-ion*) were found to be of secondary importance.

An approximate measure of the influence of an electrolyte on the flocculation of a lyophobic sol is the *critical coagulation concentration* (c.c.c.), which is the minimum concentration of that electrolyte leading to rapid coagulation under specified conditions of concentration of sol, rate of addition of electrolyte, *etc*.

Although not always a very reproducible quantity, it is nevertheless sufficient to establish the *Schultze–Hardy rule*, according to which the c.c.c is determined largely by the valency of the counter-ions. Under controlled conditions these concentrations are very roughly in the ratios 1 : 0.013 : 0.0016 for counter-ions of valency 1, 2, and 3, respectively. Other factors play a less important role. For example, the effectiveness of monovalent cations in coagulating a negatively charged sol varies in the order $Cs^+ > Rb^+ > K^+ > Na^+ > Li^+$, while for divalent ions $Ba^{2+} > Sr^{2+} > Ca^{2+} > Mg^{2+}$. The term *lyotropic series** is often applied to such sequences.

These observations clearly point to an electrostatic interpretation of the properties of aqueous lyophobic dispersions. Thus early attempts to understand coagulation by electrolytes related it to the adsorption of counter-ions and the neutralisation of the surface charge, a view supported by the empirical observation that coagulation often occurs when the zeta-potential has been reduced to some critical value around 30 mV. However, a quantitative theory has to be based on the more general concept of the electrical double layer and of the influence of electrolyte concentration on its properties.

THE DERYAGIN–LANDAU–VERWEY–OVERBEEK (DLVO) THEORY

The basic concepts of the DLVO theory were introduced in Chapter 3. We now examine the theory in more detail and compare its predictions with experimental observations.

In this case, as indicated above, the colloid stability is controlled by the form of the interaction free-energy curve as a function of particle separation. The DLVO theory in its original form considers just two contributions to this energy, namely the attractive van der Waals potential and the repulsive potential that arises when the diffuse double layers round the two particles overlap. To put this in a quantitative form we need to examine more closely the origin of the curves shown in Figures 3.6 and 3.7.

We saw in equation (3.21) that the density (ρ) of excess charge at a point in the double layer where the electrical potential is ψ is given by

$$[c(+) - c(-)]z = \rho = zc^0[\exp(-ze\psi/kT) - \exp(+ze\psi/kT)]. \tag{9.1}$$

* A similar ordering of the properties of these ions results from a consideration of phenomena such as the surface tensions of aqueous electrolytes, ionic adsorption at surfaces, ester hydrolysis, and the precipitation of albumin.

Here c^0 is the bulk concentration of a z–z valent electrolyte. However, a fundamental equation of electrostatics (*Poisson's equation*) relates this density to the way in which ψ varies with the distance from the surface:

$$\partial^2\psi/\partial x^2 = -\rho/\varepsilon, \tag{9.2}$$

where ε is the permittivity of the medium. Combining equations (9.1) and (9.2) gives the *Poisson–Boltzmann equation*, which when solved shows that at high surface potentials, but far from the surface where $ze\psi/4kT \ll 1$, the potential within the double layer falls off approximately exponentially with distance from the surface:

$$\psi = (4kT/ze)\exp(-\kappa x), \tag{9.3}$$

where, as before, $1/\kappa$ is a measure of the 'thickness of the double layer' (Figure 9.2).

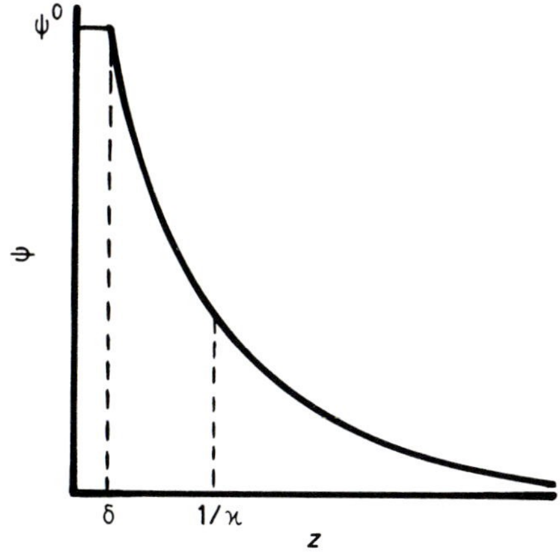

Figure 9.2 *Variation of the electrical potential as a function of the distance z from a charged surface; ψ_0 is the potential at the distance of closest approach of an ion to the surface.*

As two charged surfaces come together, the double layers overlap (Figure 9.3) and as a rough approximation the electrical potentials arising from the two surfaces (dashed curves) are additive. The electrical potential between the surfaces thus increases as shown by the full line. This increase implies an increase in the electrical contribution to the free energy of the system. It turns out that the consequent repulsive energy arising from the double-layer overlap is also exponentially dependent on the separation (H), as already sketched in Figure 3.7(b). The approximate equation for two parallel plates is, per unit area,

$$\Delta G(\text{electrostatic}) = (64c^0kT/\kappa)\exp(-\kappa H). \qquad (9.4)$$

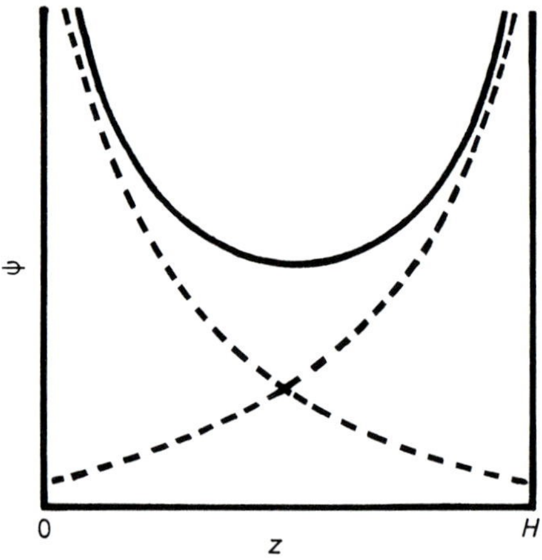

Figure 9.3 *Variation of the electrical potential between two charged surfaces a distance H apart. The total potential (full line) is the sum of the potentials arising from the two plates.*

The main concentration dependence arises from the exponential term, through the dependence of κ on concentration. When the van der Waals attractive potential is added, the total potential curve for this interaction between two parallel surfaces takes the form

$$\Delta G(\text{int}) = (64c^0kT/\kappa)\exp(-\kappa H) - A_{\text{H}}/(12\pi H^2). \qquad (9.5)$$

Typical curves illustrating the effect of electrolyte concentration on $\Delta G(\text{int})$ were shown in Figure 3.12(a). These indicate clearly the way in which an increase in electrolyte concentration leads to a lowering of the barrier to coagulation. A rough criterion for the onset of rapid coagulation is the concentration at which both $\Delta G(\text{int}) = 0$ and $\mathrm{d}\Delta G(\text{int})/\mathrm{d}H = 0$ (Figure 9.4). According to this, the critical coagulation concentration is given by

$$c^0(\text{c.c.c.}) \propto 1/(A_{\text{H}}^2 z^6), \qquad (9.6)$$

leading to the ratios 1 : 0.016 : 0.0014 for ions of valency 1, 2, and 3, in broad agreement with the experimental values quoted above. Not surprisingly it also follows from equation (9.6) that the larger the Hamaker constant A_{H} the more readily is the dispersion coagulated.

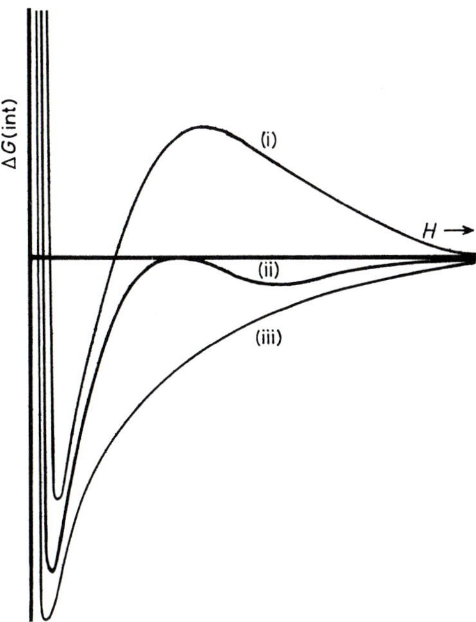

Figure 9.4 *The transition from stability* (i) *to instability* (iii) *as the electrolyte concentration is increased. The critical condition for the onset of rapid coagulation* (ii) *is that at which* $\Delta G(int)$ *at the maximum of the curve is zero.*

The above theoretical rationalisation of the Schultze–Hardy rule was claimed as an early success of the DLVO theory. However, among the approximations and assumptions made in reaching equations (9.4) and (9.6) is that the surface potential is high. On the other hand, it is known that flocculation occurs at relatively low surface potentials. In this case the simple theory suggests that the c.c.c. is proportional to z^{-2} rather than to z^{-6}. It seems possible that removal of some other approximations in the theory (*e.g.* the neglect of specific adsorption of ions) may to some extent restore agreement with the Schultze–Hardy rule.

A further prediction of the DLVO theory is that under certain conditions the total potential-energy curves may exhibit a secondary minimum. It is then expected that reversible flocculation may occur, and discussion of this follows in the next section.

The above is only a very brief account of the DLVO theory, since its full development involves rather elaborate mathematics and some necessary approximations which are probably of limited validity. Nevertheless, the general principles upon which it is based are valuable guides to an understanding of lyophobic colloids.

It is interesting to note that, if the formation of a double layer is regarded as the adsorption of counter-ions by the charged particle, then the principles concerning the effect of adsorption on interparticle forces set out in Chapter 5 can be applied. Thus, as two double layers overlap at constant potential, counter-ions are rejected from the space between the particles, the adsorption of counter-ions decreases, and, according to the general principles discussed in Chapter 5, the particles repel one another.

REVERSIBLE FLOCCULATION

In this section we examine the behaviour of systems for which the interparticle free-energy curve exhibits a shallow minimum. This may arise either in electrostatically stabilised systems under certain conditions or in sterically stabilised systems. When two particles interacting in this way come within range of their mutual attraction, they will tend to form a doublet corresponding to the separation at the secondary minimum. This tendency is opposed by the motions imparted to the particles by collisions with molecules of the dispersion medium (Brownian motion). The average energy associated with this motion is of the order of kT, so that, provided the depth of the secondary minimum is not too much greater than this, Brownian motion will keep a proportion of the particles in suspension. A state of kinetic equilibrium is set up in which the rate at which particles collide and aggregate is equal to that at which bombardment of the aggregate causes it to break up. Clearly a binary aggregate may, through further collision with a particle, form a triplet, which if its lifetime is long enough may grow to a quartet, and so on. Whether or not this growth proceeds to the formation of a macroscopic floc depends on whether subsequent additions of particles to a given cluster lead to a decrease or an increase in the free energy of the whole system. An insight into this problem may be obtained by the application of some simple thermodynamic ideas.

The problem is closely analogous to that we discussed earlier in connection with nucleation in molecular systems (Chapter 4). In both instances we are concerned with the factors which determine whether an aggregate of molecules or particles grows in size or tends to disintegrate. In this chapter we shall put the theory into a more precise form by considering the aggregation process as a sequence of addition reactions involving colloidal particles:

$$\begin{aligned} A + A &= A_2 \\ A_2 + A &= A_3 \\ &\cdots\cdots\cdots\cdots \\ A_i + A &= A_{i+1} \qquad (9.7) \\ &\cdots\cdots\cdots\cdots \\ A_{n-1} + A &= A_n \end{aligned}$$

Each step will be governed by an equilibrium constant, $K(i, i + 1)$, which, assuming that the particles behave as ideal solutes, is given by

$$K(i, i + 1) = x_{i+1}/x_i x_1, \qquad (9.8)$$

where x_i *etc.* are the mole fractions of the aggregates of various sizes. In accordance with the standard thermodynamic equation, the equilibrium constant is related to the standard free energy for the addition of a particle to an i aggregate, converting it to an $(i + 1)$ aggregate:

$$\begin{aligned} \Delta G^{\ominus}(i, i + 1) &= -kT\ln K(i, i + 1) \\ &= -kT\ln x_{i+1} + kT\ln x_i + kT\ln x_1. \qquad (9.9) \end{aligned}$$

The ratio of mole fractions of $(i + 1)$ aggregates to i aggregates is therefore given by

$$-kT\ln(x_{i+1}/x_i) = \Delta G^{\ominus}(i, i + 1) - kT\ln x_1. \qquad (9.10)$$

The standard free-energy change, $\Delta G^{\ominus}(i, i + 1)$, is determined by the potential energy of interaction of a particle with the i aggregate and is negative, while the decrease in entropy of the added particle is represented by the second term on the right-hand side. Since $\ln x_1$ is negative, this term is positive. The two terms on the right thus reflect the two contributions to the equilibrium position discussed earlier.

Equation (9.10) may be written in the form

$$x_{i+1}/x_i = x_1 \exp[-\Delta G^{\ominus}(i, i+1)/kT]. \tag{9.11}$$

If $\Delta G^{\ominus}(i, i+1)$ is not sufficiently large and negative, or if x_1 is too small, so that the right-hand side of equation (9.11) is less than unity, then $(x_{i+1}/x_i) < 1$ for all values of i. The concentration of larger aggregates will decrease with size and the equilibrium situation will be that in which the smallest aggregates and single particles predominate. The dispersion will be stable, although small aggregates may be present in kinetic equilibrium with one another and with single particles. On the other hand, if $\Delta G^{\ominus}(i, i+1)$ is large and negative, and if x_1 is sufficiently large, then $(x_{i+1}/x_i) > 1$ for all values of i, and large aggregates will form in increasing amounts and the dispersion will flocculate.

For a given set of values of $\Delta G^{\ominus}(i, i+1)$ there will be a limiting particle concentration given by $x_1 = \exp[\Delta G^{\ominus}(i, i+1)/kT]$ beyond which $(x_{i+1}/x_i) > 1$ and flocculation sets in with the formation of large flocs.

We may draw a close analogy here between the behaviour of colloidal dispersions and molecular systems. Thus the first case discussed above is analogous to the presence of clusters of molecules in a vapour approaching its condensation point, or in a solution close to saturation. The limiting concentration at which flocculation occurs corresponds to the saturation vapour pressure, or to the solubility of a solid in solution. More complex colloidal systems often exhibit phase behaviour which is paralleled by various phase separation phenomena in molecular systems. Detailed discussion of these matters is outside the scope of this book. However, pursuit of these analogies and their interpretation is a currently active area of research.

In the above discussion we have implied that $\Delta G^{\ominus}(i, i+1)$ is independent of i. In general this will not be so, but to have taken account of this variation would have complicated the argument. We shall, however, return to a more detailed discussion of equations (9.7)–(9.10) when dealing with another aggregation phenomenon, namely micellisation, in Chapter 11.

STERICALLY STABILISED SYSTEMS

In the preceding section we saw that, provided the depth of the free-energy minimum beween floc and particle [$\Delta G^{\ominus}(i, i + 1)$] is sufficiently large, gross flocculation of a weakly flocculated system occurs when the concentration of particles exceeds a *critical flocculation concentration* (c.f.c.). Conversely, because of the close relationship between floc–particle and particle–particle interactions, flocculation of a dispersion of a given concentration can be brought about by increasing the depth of the minimum in the particle–particle interaction energy. This may be the result of a decrease in the density of the adsorbed layer (Figure 3.11) or of the effective thickness as the repulsive potential becomes softer [Figure 3.12(b)]. Reduction in the range of the repulsive forces may be achieved by changing the nature of the medium, for example by adding a non-solvent for the polymer. Polymer segment–medium interactions are decreased and at some critical concentration of additive the segment–segment interactions dominate, leading to a more compact polymer layer and a reduction in δ. In bulk polymer solutions this transition from an extended chain in a 'good' solvent to a more compact configuration in a 'poor' solvent occurs at the so-called *θ-point*. One may therefore expect the concentration of added non-solvent needed to cause flocculation to correlate with that required to reach the θ-point in a bulk polymer solution. The way in which the interaction potential arising from the polymer layers varies as one passes through this θ-condition is illustrated in Figure 9.5. In a poor solvent this interaction is expected to make a contribution to the attractive potential. This point is considered in more detail in Appendix VI.

Alternatively, the θ-condition may be reached by increasing the temperature. Here the interactions between the medium and the segments of the polymer decrease because of the decreasing density of the medium, and segment–segment interactions dominate when an *upper flocculation temperature* (u.f.t.) is reached. This is analogous to the lower consolute behaviour of those polymer solutions which undergo a phase separation as the temperature is raised.

A decrease in temperature may also lead to flocculation at a *lower flocculation temperature* (l.f.t.), partly because the Brownian motion can no longer maintain the dispersed state and partly because in some instances the polymer chains 'freeze' on the surface and no longer act as a steric barrier.

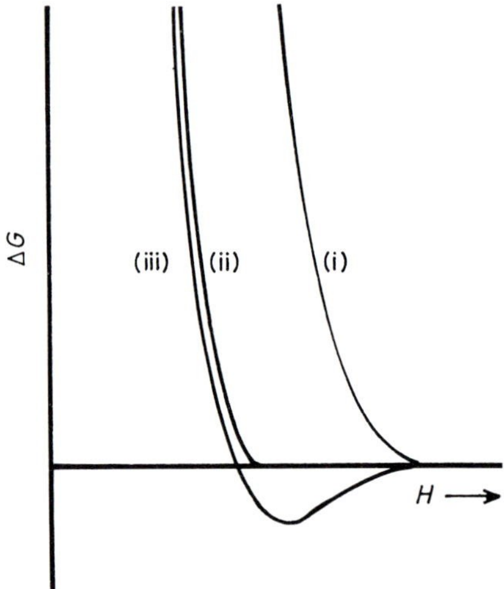

Figure 9.5 *Contribution to the interparticle potential from a polymeric surface layer in the presence of* (i) *a good solvent,* (ii) *a solvent at the θ-point (θ-solvent), and* (iii) *a poor solvent.*

In other instances the repulsive potential may become so soft [Figure 3.11(iv)] that the primary maximum is lowered and coagulation takes place into the primary minimum. This occurs if the polymer chains are too short, or not present at sufficient density on the surface to provide an adequate steric barrier. In this case it is often found that there are no conditions under which the dispersion is stable.

Interaction between the polymer layers is not the only factor influencing flocculation. Thus changing the medium also affects the effective Hamaker constant, A_H. It is then the increased attractive potential arising from an increase in A_H that may determine the flocculation. In this case there may be little correlation with the θ-point.

BRIDGING FLOCCULATION

If a solution of a high-molecular-weight polymer is added to a dilute dispersion, *bridging flocculation* may occur. In this it is supposed that the two ends of a polymer chain adsorb on separate particles and draw them together (Figure 9.6). This phenomenon is exploited in water purification where the addition of a few parts per million of a high-molecular-weight polyacrylamide leads to flocculation of remaining particulate matter in the water.

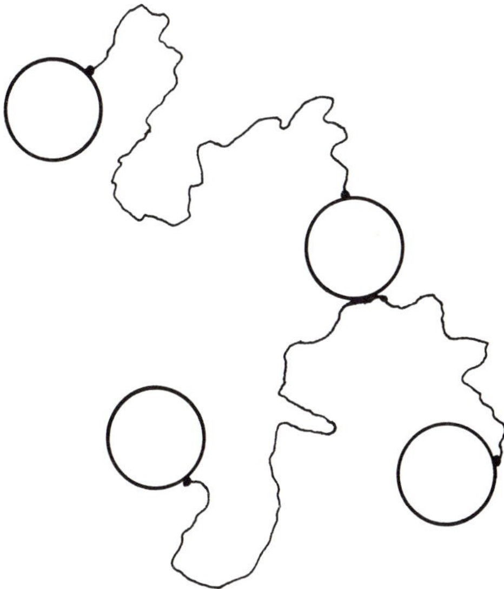

Figure 9.6 *The origin of bridging flocculation. This depends on the presence of a high-molecular-weight polymer in a good solvent. The ends, or widely separated segments, of the polymer are adsorbed on different particles, drawing them together.*

DEPLETION FLOCCULATION

Another effect arises if a non-adsorbed or weakly adsorbed polymer is added to the dispersion. When two particles come close together, polymer is excluded from the space between them: in effect the polymer is negatively adsorbed. If the volume of solution from which polymer is excluded is V^x, the negative adsorption is $\Gamma = -c_2 V^x$, where c_2 is the bulk concentration of polymer. As the particles come together, V^x decreases (Figure 9.7) and the adsorption becomes less negative. Referring to the general principles outlined in Chapter 5 (pages 66 and 70), we see that

$$\mathrm{d}\Gamma/\mathrm{d}H = -c_2(\mathrm{d}V^x/\mathrm{d}H) = -c_2 A, \qquad (9.12)$$

which is negative. Here A is the area of the excluded zone. Consequently there is an attraction between the particles. The attractive force may be calculated by writing equation (5.12) in its integral form (putting $\Gamma = 2\Gamma_2^{(1)}$):

$$\mathcal{F}(H) = \int_{\mu_2=-\infty}^{\mu_2} (\mathrm{d}\Gamma/\mathrm{d}H)\mathrm{d}\mu_2. \tag{9.13}$$

If we make the gross approximation of assuming that the solution is ideal, *i.e.* $\mathrm{d}\mu_2 = RT\mathrm{d}\ln c_2$, then

$$\mathcal{F}(H) = -RTA\int_0^{c_2} \mathrm{d}c = -c_2RTA. \tag{9.14}$$

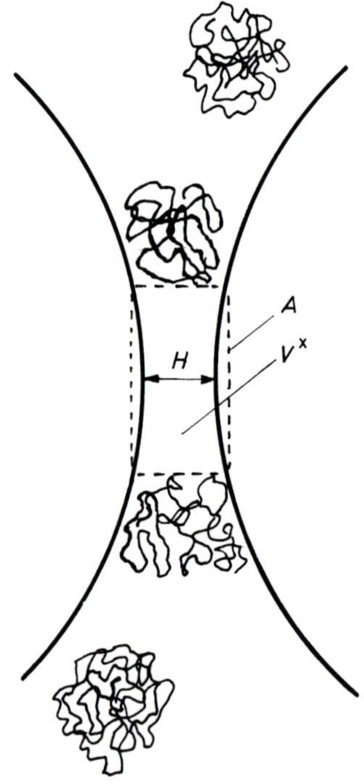

Figure 9.7 *Origin of depletion flocculation. Polymer molecules are excluded from a volume V^x and an area A between the particles at a separation of H. In this region there is a negative adsorption given by $-c_2V^x$. As H decreases, the adsorption becomes less negative (i.e. increases) and an attractive force results.*

An alternative way of looking at this phenomenon is to say that the gap between the particles prevents polymer from entering the space between them and so acts like an osmotic membrane permeable only to the solvent. Solvent thus tends to diffuse into the bulk, reducing the amount of solvent between the surfaces. This is driven by an osmotic pressure Π, which acting over an area A creates a force $-\Pi A$. Again, assuming the solution to be ideal [*cf.* equation (6.25)],

$$\mathcal{F}(H) = -\Pi A = -c_2 RTA, \qquad (9.15)$$

which is the same conclusion as that reached by the earlier argument. Of course, a more rigorous theory must take into account the strong non-ideality of polymer solutions, but this again shows the equivalence of the two approaches.

At this point we see qualitatively that the mechanisms of steric stabilisation and depletion flocculation are closely related. In the former instance the concentration of polymer segments in the space between the particles increases as the particles come together, leading to a repulsion caused by the osmotic flow of solvent into this space; in the latter case the concentration between the particles is lower than that in the bulk, and diffusion of solvent out of the interparticle space results in an attraction.

KINETICS OF COAGULATION

The discussion of the preceding sections leads immediately to a consideration of the kinetics of coagulation. Two cases are important, depending on whether or not coagulation is inhibited by an energy barrier.

In the absence of a barrier to coagulation, and if the primary minimum is deep, every collision between a particle and a floc will lead to the growth of the floc. The rate of coagulation is then controlled entirely by the kinetics of the diffusion process leading to particle–particle collision. The theory of *fast coagulation* was developed originally by Smoluchowski (1918) and elaborated by Müller (1926). The rate equation has the same form as that for a bimolecular reaction:

$$\mathrm{d}\nu/\mathrm{d}t = -8\pi RD\nu^2, \qquad (9.16)$$

where ν is the concentration of single particles and $\mathrm{d}\nu/\mathrm{d}t$ is the rate at which they disappear. R, the collision diameter, is twice the particle radius and D is the diffusion coefficient for single particles. The time for reduction of the initial concentration by a factor of two (the *half-life*) is thus

$$t_{1/2} = \frac{1}{(8\pi R D \nu_0)}, \qquad (9.17)$$

where ν_0 is the initial particle concentration. Making use of the equation for the diffusion coefficient (6.10), we find that in water at 25 °C

$$t_{1/2}/\mathrm{s} = \frac{3\eta}{4kT\nu_0} = 2 \times 10^{11}/(\nu_0/\mathrm{cm}^{-3}). \qquad (9.18)$$

At a concentration of 10^{10} particles cm^{-3} (or $10^{16}\,\mathrm{m}^{-3}$) the half-life is about 20 s.

This theory takes account only of collisions between single particles and neglects the collisions of single particles with aggregates and of aggregates with one another. It applies, therefore, strictly only to the early stages of coagulation. When such effects are taken into account, it is found that the process still follows overall bimolecular kinetics, but ν in equation (9.16) is now the total number of particles irrespective of their size and $t_{1/2}$ now refers to the time at which the total number of particles has been halved.

If the adherence between two particles which collide is inhibited by an energy barrier, the process resembles a bimolecular chemical reaction involving an activation energy. In this case we have to deal with *slow coagulation*. On the basis of this analogy one might expect therefore that

$$\mathrm{d}\nu/\mathrm{d}t = -8\pi R D \nu^2 \exp(-V_{\max}/kT), \qquad (9.19)$$

where $V_{\max}$ is the height of the free-energy barrier preventing coagulation. The coagulation is slowed down by a factor W, called the *stability ratio*, which, dividing equation (9.16) by (9.19), is given by

$$W = \exp(V_{\max}/kT). \qquad (9.20)$$

However, the analogy is not quite exact, for two main reasons. First, because the range of action of colloidal particles is so much longer than the chemical (electronic) interactions involved in forming an activated complex in a chemical reaction, the last stage of diffusion of two particles toward one another takes place against the repulsive force between the two particles, thus reducing the collision rate. A more exact equation for the stability ratio is

$$W = R\int_{r=2R}^{r=\infty}(1/r^2)\exp(V/kT)\mathrm{d}r, \tag{9.21}$$

where V is the potential when the centres of the particles are separated by a distance r.

In the case of charged particles an approximate equation is

$$W = (1/\kappa R)\exp(V_{\max}/kT) \tag{9.22}$$

or

$$\ln W = -\ln\kappa R + V_{\max}/kT. \tag{9.23}$$

It turns out that $V_{\max}$ is a roughly linear function of $\ln c^0$, where c^0 is the electrolyte concentration, so that graphs of $\ln W$ against $\ln c^0$ are approximately linear, extrapolating to $\ln W = 0$ at the critical coagulation concentration. This prediction is borne out experimentally.

A second factor reducing the rate of collision arises from the fact that, as two particles approach closely, liquid has to flow out from the region between them. Experimental evidence based on measurements of rapid coagulation (which is similarly affected by this hydrodynamic effect) indicates that the collision rate is reduced to about half its expected value.

Coagulation which occurs under the influence of Brownian motion is called *perikinetic coagulation*. However, the rate of particle collision can be enhanced and the rate of coagulation increased by the hydrodynamic forces created by rapid stirring, or flow through a tube at high shear rates. This is called *orthokinetic coagulation*. The hydrodynamic effects increase with particle size and become dominant for particles larger than a few micrometres. This means that if the initial particles are smaller than this then the early stages of coagulation are controlled by Brownian motion, but as the flocs increase in size their rate of growth can be accelerated by stirring or shaking. It is a common observation that the relatively diffuse flocs produced when coagulation occurs under quiescent conditions can be greatly concentrated by shaking.

HETEROCOAGULATION

Charge-stabilised dispersions are, as we have seen, coagulated by the adsorption of counter-ions in the electrical double layer. Even more effective are charged particles of opposite sign. Thus *co-coagulation* or *heterocoagulation* occurs when negatively charged and positively charged dispersions are mixed. One interesting historical example is the preparation of Purple of Cassius. In this a gold chloride solution is reduced by stannous chloride. The gold particles so produced are negatively charged, while the stannic ions resulting from the reduction form colloidal stannic hydroxide, which is positively charged. The oppositely charged particles form a heterocoagulum which precipitates as a loose purple floc. When dried and fused with glass, this produces ruby glass.

This phenomenon is quite general and has a number of important practical applications.

STRUCTURE OF FLOCS AND SEDIMENTS

A factor of very great importance in the handling of colloidal dispersions is the structure of flocs and sediments. Thus, in the filtration of a flocculated suspension it is necessary to have as open a structure as possible so that liquid can flow through the filter cake.

The morphology of flocs is determined to a large extent by the nature of the interparticle forces involved. Thus, if the particles are strongly repelling but sediment under gravity or centrifugation, they tend to form a close-packed structure because the particles can slide across one another under the influence of Brownian motion and seek out a configuration of minimum

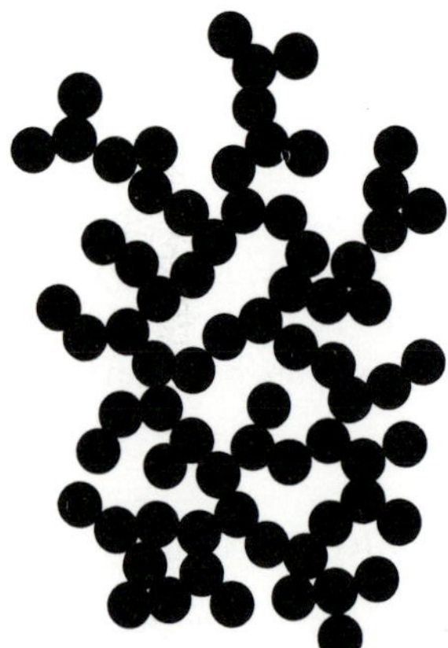

Figure 9.8 *Open structure formed by a floc produced by strongly attracting particles.*

energy. Figure 1.3 showed an example of the ordered structures formed when a stable polymer latex dispersion is allowed to dry slowly.

On the other hand, if the dispersion is flocculated, particles on contact tend to stick in the configuration in which they collide and thus tend to form a much more open structure (Figure 9.8). Of particular practical interest are weakly flocculated dispersions because, although they may sediment or cream, they form relatively open structures that are readily dispersed by shaking or stirring.

The formation of a close-packed structure or 'clay' on storage of a dispersion has to be avoided in many technical applications since such sediments are difficult to re-disperse (see Chapter 14).

Chapter 10

How are Colloidal Dispersions Destroyed? II Coalescence and Particle Growth

INTRODUCTION

In the preceding chapter we saw how colloidal dispersions can be destroyed by aggregation processes in which a reduction of the free energy of the system occurs while the surface area of the individual particles remains constant [see equation (2.3)]. In the present chapter we investigate systems in which the lowering of free energy is achieved by a reduction of interfacial area, the interfacial tensions being relatively unchanged [see equation (2.4)]. Three main mechanisms may be involved: sintering or particle coalescence, particle growth or ripening, and droplet coalescence.

SINTERING OR PARTICLE COALESCENCE

Let us consider first a fine powder of colloidal dimensions produced by comminution. To every fracture surface of each particle there must be somewhere a matching surface on some other particle (Figure 10.1) such that if these two particles were brought together in the right orientation they would match and, without any activation energy, would coalesce to form a larger particle [Figure 10.1(c)]. In effect we are considering the reverse of the breaking process discussed in Chapter 2. The situation depicted in Figure 10.1 would arise from one chance encounter among millions between two particles. This is exceedingly improbable and a few simple calculations show that the chances of eliminating the colloidal state by reconstructing the original solid – as though it were a giant three-dimensional jigsaw puzzle – are so small that, if this were the only mechanism available for reducing the surface area, then a finely divided solid of that kind would remain stable indefinitely – probably for a period of time at least equal to the age of the universe!

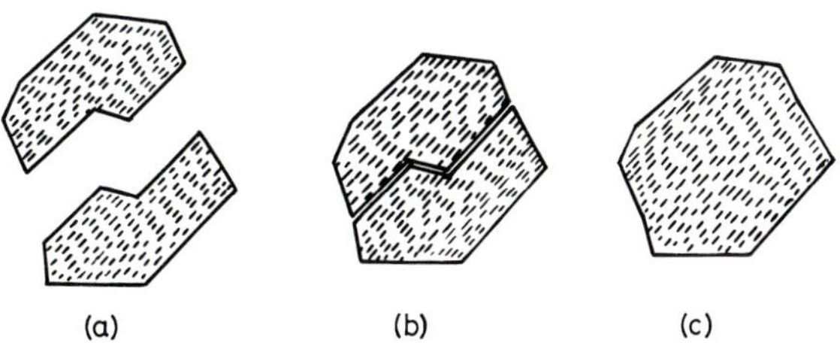

Figure 10.1 *Coalescence of two solid particles with an exact match of crystal lattices.*

We must, therefore, consider alternative mechanisms of particle coalescence such as that shown in Figure 10.2. Here we consider any two particles coming into contact in a random orientation [Figure 10.2(b)]. The combination of these two particles to form one larger particle of reduced surface area can occur only if the surface molecules are able to rearrange themselves to provide some kind of match (through a grain boundary) between the crystal lattices of the two particles [Figure 10.2(c)]. The activation energy for the displacement of an atom from its lattice position is large and can be acquired only if the temperature is raised towards the melting point of the solid. It is often found that when the temperature reaches about two thirds of the melting temperature (in Kelvins) – the so-called *Tamman temperature* – atomic mobility increases and particle fusion by *sintering* can take place. The activation energy for this is essentially that for the diffusion of atoms (or molecules) in the solid lattice adjacent to the surface. The exact mechanism of sintering depends on the system concerned. The sintering of solid particles often seems to proceed by the formation of 'necks' between touching particles (Figure 10.3). In the case of solids having an appreciable vapour pressure, sintering may be assisted by transport through the vapour phase, while the application of pressure to the packing often favours sintering. It is also clear that sintering is possible only in the absence of a stabilising agent, which would introduce a repulsion between the surfaces.

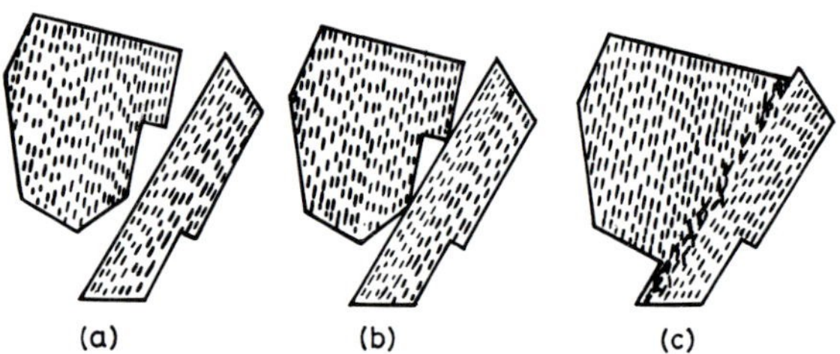

Figure 10.2 *Coalescence of two solid particles by formation of a grain boundary.*

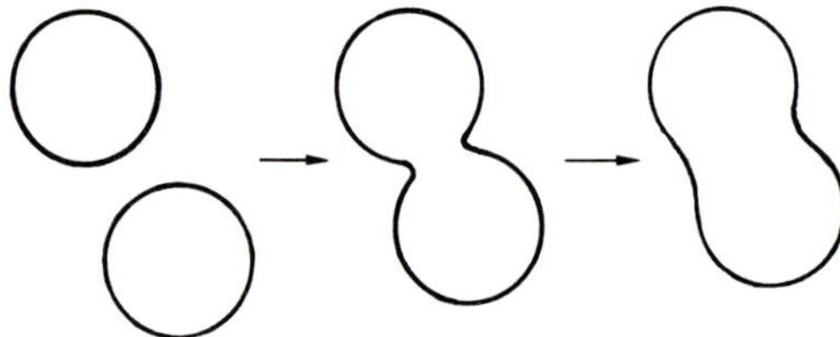

Figure 10.3 *Coalescence of two solid particles by sintering.*

The sintering process is of particular importance in powder technology, although here the particles are usually somewhat larger ($> 30\ \mu m$) than the conventional colloid range.

A clearer fundamental understanding of sintering obviously depends upon a more complete knowledge of the surface tension of solids.

PARTICLE GROWTH BY 'RIPENING'

If the solid particles are slightly soluble in the dispersion medium (or, if the medium is a gas, the solid is slightly volatile), then an alternative mechanism is available by which the system can reduce its surface area without the particles having to collide. Because of the phenomenon of surface tension the solubility of very small particles (or their vapour pressures) is greater than that of a large particle [see equations (5.23) and (5.24)]. Consequently, if particles of different sizes are present, there will be a tendency for molecules to dissolve (or evaporate) from the smaller particles (having a higher solubility or vapour pressure) and to precipitate (or condense) on the larger ones. Smaller particles tend to be eliminated, and the average particle size increases indefinitely. The rate of this process – called *Ostwald ripening* – decreases as the particles grow, since the solubility or vapour pressure difference between larger particles of slightly different sizes is less than for very fine particles differing in size by the same amount. Furthermore, the distance over which molecules have to diffuse increases as the number of particles in a given volume decreases. In addition, one expects the particle size distribution to become narrower as the particles grow. The activation energy barrier to particle growth in this case arises from the energy needed to detach molecules from the solid, *i.e.* is related to the energy of solution or of evaporation, while the rate of growth is also controlled by molecular diffusion through the medium. The higher the solubility (lower energy of solution) or the vapour pressure (lower energy of evaporation) the more rapidly is ripening likely to occur. Indeed in some instances it may occur so rapidly that the system cannot be maintained in the colloidal state.

Although colloidal systems may be in a metastable state in relation to adhesion between the particles, it is important to stress that, if they have any significant solubility or volatility, they are in general thermodynamically unstable with respect to particle growth processes.

However, there are special circumstances under which such a colloid may be effectively stable, despite the fact that the free energy of a particle decreases as its particle size increases. Thus if the particle concentration is high, and the concentration of solute low, then growth of the particles may deplete the solution to below the solubility of the particles, and particle growth will be inhibited. An analogous process may well play a significant role in, for example, the formation of stable rain drops in a cloud.

Ostwald ripening is of considerable importance in several technological situations. For example, in the preparation of photographic 'emulsions' the silver halide particles are subjected to carefully controlled ripening. On the other hand, in agrochemical preparations it is important that ripening should not occur otherwise the particle size of, for example, pesticide components would increase seriously in storage.

DROPLET COALESCENCE

In the case of emulsions the process of droplet coalescence plays a role akin to sintering in solids, although the mechanism is somewhat different. Here the successive stages, shown in Figure 10.4, involve the approach of the droplets, the thinning of the layer of dispersion medium between them to a very thin film [Figure 10.4(b)], and finally the bursting of this film to allow the material in the two drops to combine to form a larger drop of lower surface area; coalescence of two drops of equal size leads to a 41% reduction in surface area [Figure 10.4(c)]. In this case the main barrier opposing the process is that associated with the final stages of the thinning and bursting of the film separating the drops (Figure 10.5). These final stages involve film thicknesses in the colloidal size range, so that, even when the emulsion droplets themselves are larger than the upper limit of the colloid range, their stability to coalescence depends very largely on the colloidal properties of the film between them, including its rheological behaviour.

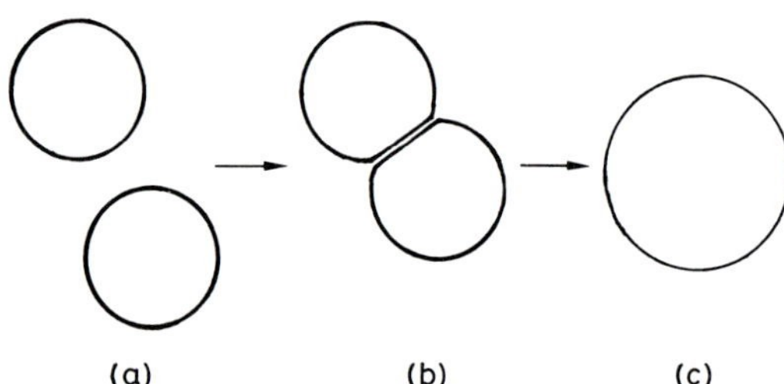

Figure 10.4 *Coalescence of two emulsion droplets involving the thinning of the film between them and then formation of a larger drop.*

The process of film thinning in the case of emulsion droplets differs in an important respect from that in the approach of two solid particles. In both cases the intervening liquid has to flow out of the space between the approaching bodies. However, if the emulsion is in a medium of high viscosity, the liquid/liquid interfaces may be deformed, leading to the situation depicted in Figure 10.6. A lens of liquid becomes trapped between the droplets and its elimination may be a slow process which inhibits the eventual coalescence.

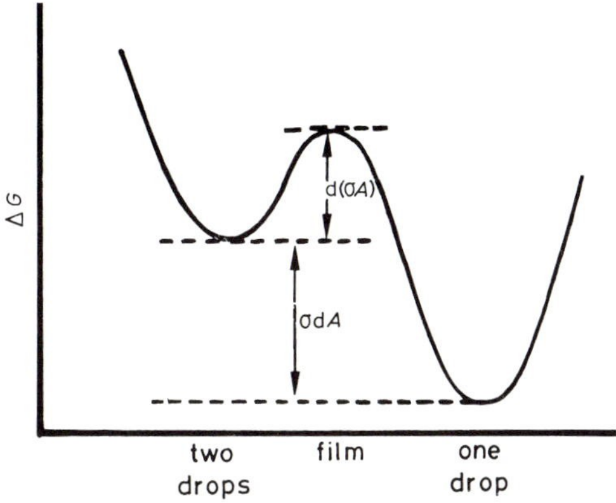

Figure 10.5 *Variation of the free energy during the coalescence of two emulsion droplets. During the thinning of the film between them there is a change in both the interfacial area and the film tension, leading to an increase in free energy. After coalescence the surface tension returns to its original value but the surface area has decreased.*

Figure 10.6 *If the fluid between the droplets has a high viscosity, a lens of liquid may be trapped. Expulsion of this to form a thin film may be a slow process.*

The stability of emulsions thus depends on the resistance to film thinning and rupture, which we shall consider in greater detail in Chapter 12. Meanwhile it is clear that practical problems of de-emulsification will involve the identification of circumstances which will reduce this resistance.

Processes akin to ripening can of course also contribute to droplet growth in emulsions, although in most cases it is not the dominant mechanism.

Analogous considerations apply to the destruction of foams in which the film of liquid lies between two gas bubbles rather than between two liquid droplets.

Chapter 11

Association Colloids and Self-Assembly Systems

INTRODUCTION

The formation of colloidal dispersions by growth from the molecular state is often controlled by a nucleation step. An embryo has to achieve a certain minimum size before growth can proceed spontaneously (Chapter 4) but is then limited by the availability of material in the bulk phase from which the particle is growing.

There is, however, a large and important class of colloids in which nucleation is absent. Growth is spontaneous, but the structures so formed are usually limited by geometric and energy factors to a finite size often towards the lower end of the colloid size range. This class comprises *association colloids*, or in more general terms *self-assembly systems*: it includes not only micelles but many more complex forms, *e.g.* vesicles with, as extreme examples, biological structures such as cell membranes.

This chapter outlines the main features of systems of this kind and indicates in broad terms the factors leading to their formation and controlling their size and shape. Discussion of another group of spontaneously formed colloids – *gels* – will be deferred to Chapter 13.

MICELLISATION

The physical properties of dilute solutions of certain surface-active substances, when plotted as a function of solution concentration, show a more or less sharp break, which occurs at roughly the same point for different properties. Typical examples are the breaks in the curves of surface tension, electrical conductivity, and osmotic pressure, as illustrated schematically in Figure 11.1. The existence of this phenomenon is attributed to the formation, at a certain concentration, of aggregates of surfactant molecules, called *micelles*, whose contribution to the physical properties of the system is different from that of individual molecules. The solution concentration at which this occurs is called the *critical micellisation concentration* (c.m.c.). Evidence from a variety of sources shows that the micelles initially formed usually contain a relatively small number of molecules (50—100) and are roughly spherical in shape; at higher concentrations and under appropriate conditions these spherical micelles may adopt a disc-like, cylindrical, ellipsoidal, or laminar form (Figure 11.2).

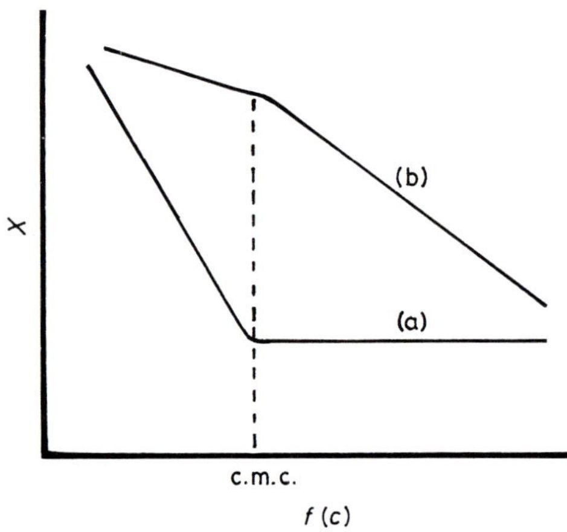

Figure 11.1 *Variation of the properties of a solution of a surfactant with concentration c, illustrating the onset of micellisation:* (a) *surface tension against lnc,* (b) *molar conductivity against* $c^{1/2}$.

We observe that surface-active substances that form micelles possess one common feature, namely they are *amphipathic*, which means that the molecule consists of two parts, one of which is highly soluble in the medium concerned and the other insoluble. We shall deal mainly with aqueous solutions, although it is important to realise that micellisation is not confined to water as solvent but can occur in many non-polar media. In the case of aqueous systems the surfactant molecule consists of a hydrophilic group (*head group*) to which is attached a hydrophobic hydrocarbon group (*tail*).

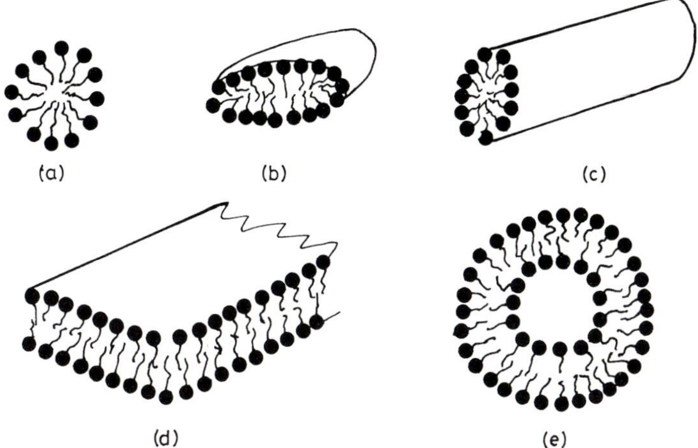

Figure 11.2 *Some types of micellar structure:* (a) *spherical,* (b) *disc-like,* (c) *cylindrical,* (d) *lamellar,* (e) *spherical vesicle. The head groups are shown as filled circles.*

Two main types of such compounds may be distinguished: *ionic* (including *anionic, cationic,* and *ampholytic*) and *non-ionic.* Ionic surfactants possess a hydrophilic head group such as $—OSO_3^-$, while an important class of non-ionics are block copolymers of the type A_nB_m, where the A-chains are hydrophilic (*i.e.* the homopolymer A_n is water soluble) and the B-chains are hydrophobic or insoluble in water. With ampholytic surfactants the nature of the ionised group depends on pH. Some typical examples of surfactants that form micelles are listed in Table 11.1.

MECHANISM OF MICELLISATION

The behaviour of a solution of amphipathic molecules reflects the opposing tendencies of one part of the molecule to separate out as a distinct phase while the other tends to stay in solution. The factors that have to be considered in discussing the effects of this balance therefore include (i) the interactions of hydrocarbon chains with water, (ii) the interaction of hydrocarbon chains with themselves, (iii) the solvation of the head group, and (iv) the interaction between the solvated head groups mediated in the case of ionic groups by their ionic atmospheres. The balance will clearly also be influenced by the relative sizes of the hydrophilic

Table 11.1 *Some typical micelle-forming surfactants.*

Type and general formula (R_n is an alkyl tail group)	*Examples*	*Head group*	*Approx. c.m.c./mol dm^{-3} at 25 °C (aggreg. no. in parens.)*
Anionic			
R_nCOONa	Na hexadecanoate	$—COO^-$	1×10^{-3}
R_nOSO_3Na	Na dodecyl sulphate (SDS)	$—OSO_3^-$	8.1×10^{-3} (58)
R_nOOCCH_2	'Aerosols'	$—SO_3^-$	$n = 6$, 1.7×10^{-2} $n = 8$, 1.6×10^{-3}
$R_nOOCCHSO_3Na$			
Cationic			
R_nNH_3Cl	Dodecylammonium chloride	$—\overset{+}{N}H_3$	1.3×10^{-2} (56)
R_nNR_3Br	Decyltrimethylammonium bromide	$—\overset{+}{N}Me_3$	6.8×10^{-2} (36)
$R_nN(C_5H_5)I$	Dodecylpyridinium iodide Cetylpyridinium chloride (CPC)	$—\overset{+}{N}(C_5H_5)$	5.6×10^{-3} (87) 9.4×10^{-4}
Ampholytic			
$R_n\overset{+}{N}H_2COO^-$	Alkyl betaines	$—\overset{+}{N}H_2COOH$ or $—NHCOO^-$	$n = 10$, 1.8×10^{-2} $n = 12$, 1.8×10^{-3} $n = 14$, 1.8×10^{-4} $n = 16$, 1.8×10^{-5}
Non-ionic			
$R_n(OCH_2CH_2)_mOH$	Polyethylene oxides ('C_nE_m')	$—(OCH_2CH_2)_mOH$	$\left.\begin{matrix} m = 6, \\ n = 8, \end{matrix}\right\} 9.9 \times 10^{-3}$ $n = 10$, 9.9×10^{-4} $n = 12$, 8.7×10^{-5}

Table 11.1 *(cont.)*

Type and general formula (R_n is an alkyl tail group)	*Examples*	*Head group*	*Approx. c.m.c./mol dm^{-3} at 25 °C*
Fatty (n) acid esters of anhydrosorbitols (mixture of isomers)	'Spans'	e.g. 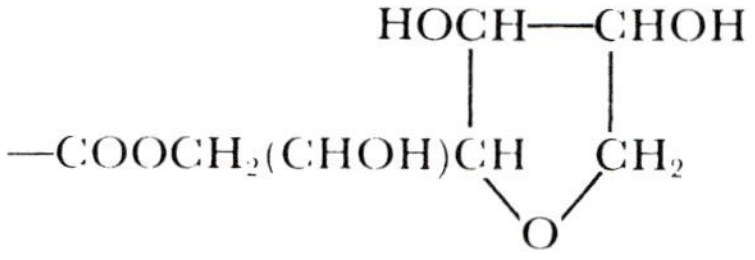 HOCH—CHOH │ │ —$COOCH_2$(CHOH)CH CH_2 \ / O	Oil soluble $n = 15$, 1.5×10^{-2} in C_6H_6
Fatty (n) acid esters and ethylene oxide (EO) condensates of anhydrosorbitols (mixture of isomers)	'Tweens'	e.g. 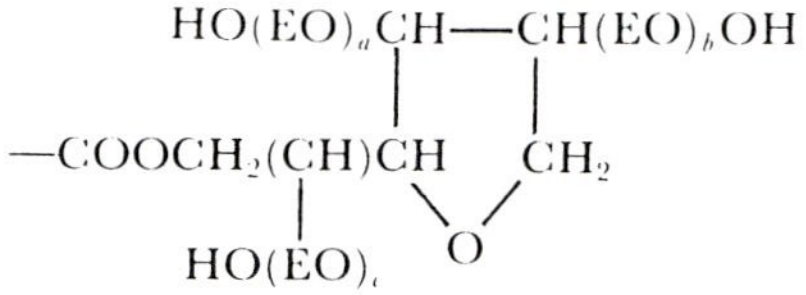 $HO(EO)_a$CH—CH$(EO)_b$OH │ │ —$COOCH_2$(CH)CH CH_2 │ \ / $HO(EO)_c$ O $a + b + c = m$	$m = 5$—25 $n = 11, 13, 15, 17$ $m = 20$, $n = 11$ } 8×10^{-3}
Polymeric			
Polyvinyl alcohol			
Methyl cellulose			
Fluorocarbon surfactants			
$C_nF_{2n+1}COOK$	K perfluoro-octanoate	$—COO^-$	2.9×10^{-2}

and hydrophobic portions of the molecule, *i.e.* by its *hydrophilic–lipophilic balance* (HLB). For example, it is interesting to observe that in several of the series listed in Table 11.1 an increase in the length of the hydrocarbon chain by two methylene groups reduces the c.m.c. by a factor of about ten. Account must also be taken of geometrical and packing factors which have to be satisfied, in addition to the above energetic factors.

The problem is that of seeking a structure which will minimise the area of contact between hydrocarbon chains and water while maintaining maximum interaction between the hydrophilic moiety and water. One such structure is a spherical micelle [Figure 11.2(a)], which must, however, satisfy the appropriate geometrical constraints. In the first place, the radius of the sphere must be of the same order as the length of the hydrocarbon chain. If it were larger, then at least some of the head groups would have to be embedded in the hydrocarbon core; if it were much less, then the hydrocarbon/water contacts would increase. Secondly, if the hydrocarbon chains are to be accommodated in the core at roughly the same density as that of a liquid hydrocarbon, then the head groups cannot be close-packed on the surface of the sphere, but at least some of the surface must expose hydrocarbon groups. A simple calculation suggests that for a spherical micelle not more than one third of the surface is occupied by head groups. This expectation is confirmed by neutron-scattering studies (Figure 11.3). Finally, the tendency of the area of contact between hydrocarbon and water to contract must be balanced by the repulsion between the head groups, which tends to increase the area.

Merely listing the factors controlling micellisation shows that any valid theory is bound to be complicated in detail. However, some useful generalisations can be developed from first principles.

We have already seen in Chapter 9 that the aggregation of particles can be discussed in terms of a sequence of addition reactions, shown in equation (9.7). Exactly the same arguments can be applied to the case of micellisation, but we have to discuss the problem in a little more detail since, unlike flocs, micelles do not grow to a macroscopic size. In particular we have to take account of the fact that the successive equilibrium constants $K(i, i + 1)$ [equation (9.8)] depend on i.

Instead of identifying the entities in equation (9.7) with colloid particles, we now take them to be individual surfactant molecules.

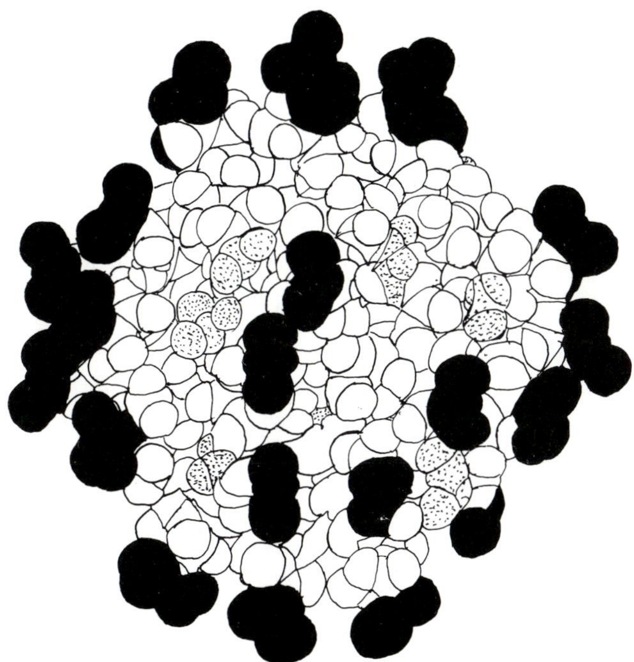

Figure 11.3 *Computer-generated space-filling model of a spherical micelle of dodecanate, which is in good agreement with low-angle neutron-scattering measurements. Black: head groups; white: hydrocarbon chains; stippled: terminal methyl groups. A substantial proportion of the outer surface is occupied by hydrocarbon groups in contact with the solvent.*
(Sketched from K. A. Dill *et al.*, *Nature*, 1984, **309**, 42)

For our present purpose it is convenient to consider the overall equilibrium constant for the formation of an aggregate containing i monomers:

$$iA = A_i, \tag{11.1}$$

for which

$$K_i = x_i/(x_1)^i. \tag{11.2}$$

The corresponding standard free energy for the formation of one mole of the i-mer is given by

$$\Delta G^{\ominus}(i) = -RT\ln K_i = -RT\ln x_i + iRT\ln x_1, \tag{11.3}$$

while the average standard free-energy change per mole of

monomer is

$$\Delta G^{\ominus}(i)/i = -(RT/i)\ln x_i + RT\ln x_1. \tag{11.4}$$

At the c.m.c. one size of micelle predominates, for which i is 50—100, while x_i becomes of the same order as x_1. The first term on the right-hand side of equation (11.4) can therefore be neglected, and an approximate expression for the standard free-energy change of micellisation *per mole of surfactant* is

$$\Delta G^{\ominus}(i)/i = RT\ln x_1(\text{c.m.c.}). \tag{11.5}$$

The problem of calculating the distribution of micelle sizes reduces to that of establishing the dependence of $\Delta G^{\ominus}(i)$ on i. Since there is good evidence that the equilibrium mixture at and above the c.m.c. contains only a low concentration of species other than those of a size close to the mean micelle size, we must seek a form of $\Delta G^{\ominus}(i)$ which leads to this result. Some early theories assumed that the standard free energy is a linear function of i. This would mean that $\Delta G^{\ominus}(i, i + 1)$, which is equal to $\Delta G^{\ominus}(i + 1) - \Delta G^{\ominus}(i)$, is constant. However, we saw in Chapter 9 that, if this is so, above a certain concentration aggregates will grow to macroscopic size, contrary to the limiting sizes associated with micellisation. Other theories have set $\Delta G^{\ominus}(i)$ equal to zero for all values of i except for a particular value of i at the c.m.c. This implies that micellisation occurs by the simultaneous association of i monomer molecules, which is physically unrealistic.

To explain micellisation it is necessary for the curve of $\Delta G^{\ominus}(i)$ as a function of i to have a more complicated form, such as that sketched in Figure 11.4. An equation leading to a curve of this kind, and which has a plausible theoretical basis that we shall discuss below, is the following:

$$\Delta G^{\ominus}(i)/RT = -a(i - 1) + b(i - 1)^{2/3} + c(i - 1)^{4/3}. \tag{11.6}$$

A suitable choice of parameters is $a = 18.00$, $b = 19.57$, and $c = 1.227$. That this equation provides an acceptable representation of the micellisation phenomenon may be shown in the following way.

We write equation (9.10) in the form:

$$\ln(x_{i+1}/x_i) = \ln x_1 - \Delta G^{\ominus}(i, i + 1)/RT. \tag{11.7}$$

Now $\Delta G^{\ominus}(i, i + 1)$ is just the slope of the curve of $\Delta G^{\ominus}(i)$ as a function of i, so that

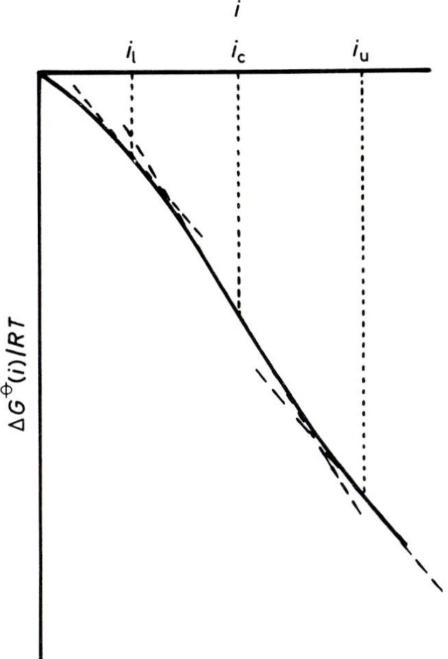

Figure 11.4 *Free energy of formation of a micelle of size i as a function of i; i_c is the value of i at which the curve has a point of inflexion, and i_l and i_u are the values of i at which the curve has a slope of $\ln x_1$.*

$$\ln(x_{i+1}/x_i) = \ln x_1 - (1/RT)[\mathrm{d}\Delta G^{\ominus}(i)/\mathrm{d}i]. \qquad (11.8)$$

Whether or not x_i increases or decreases with increase in i is therefore determined by the relative values of $\ln x_1$ and the slope of the curve of $\Delta G^{\ominus}(i)/RT$ against i: if

$$(1/RT)[\mathrm{d}\Delta G^{\ominus}(i)/\mathrm{d}i] < \ln x_1 \text{ then } x_{i+1} > x_i, \qquad (11.9a)$$

while if

$$(1/RT)[\mathrm{d}\Delta G^{\ominus}(i)/\mathrm{d}i] > \ln x_1 \text{ then } x_{i+1} < x_i. \qquad (11.9b)$$

At very low concentrations, when $\ln x_1$ is large and negative (say $|\ln x_1| > 9$), then at all points on the curve $(1/RT)[\mathrm{d}\Delta G^{\ominus}(i)/\mathrm{d}i]$ is greater than (*i.e.* less negative than) $\ln x_1$, and consequently x_i decreases with i as shown by curve (i) in Figure 11.5; only monomeric surfactant is present in appreciable amounts. As x_1 increases and $\ln x_1$ becomes less negative, a concentration is reached ($\ln x_1 = -8.57$ in this numerical example) at which $\ln x_1$ is just equal to the slope of the free-energy curve at the point of inflection C. At this point $x_{i+1} = x_i$, but for both higher and lower values of i x_i falls off with increase in i, as shown by curve (ii) in Figure 11.5.

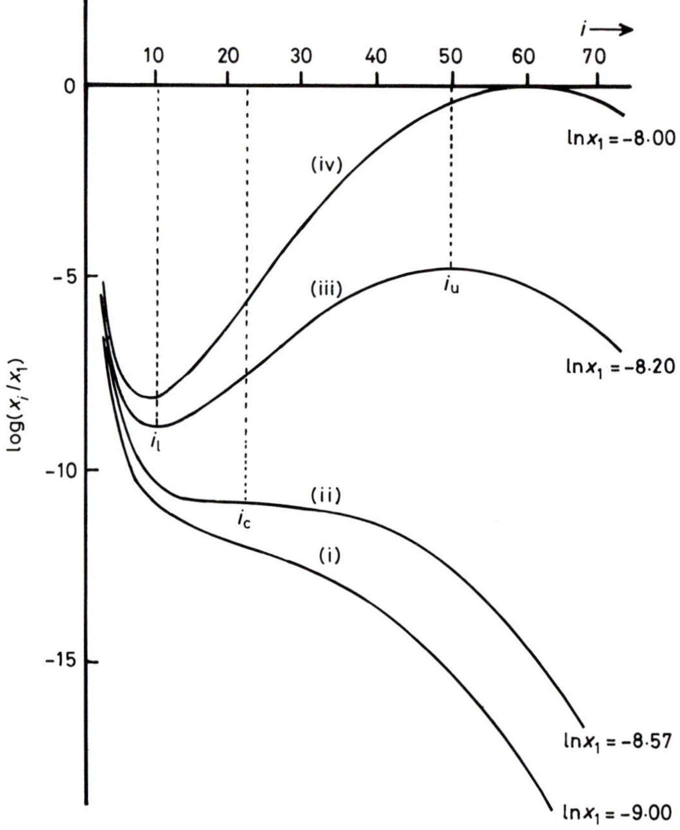

Figure 11.5 *Logarithm of the ratio of the mole fraction of micelles of size i to that of monomer as a function of i for various monomer concentrations according to equation (11.6) and the parameters given in the text;* $\ln x_1 = -8.00$ *corresponds to* $x_1 = 3.4 \times 10^{-4}$.

On further increase in $\ln x_1$ (say to -8.20), lines of this slope are tangential to $\Delta G^{\ominus}(i)/RT$ at two points i_l and i_u (Figure 11.4), and at these points $d(\ln x_i)/di = 0$. However, it is seen that the curve of $\ln x_i$ against i has a minimum at i_l and a maximum at i_u [Figure 11.5, curve (iii)]. But the concentrations of all aggregates are still negligible compared with the concentration of monomers. When $\ln x_1$ rises to -8.00, aggregates with i in the range 40—80 are all present at greater than 10% of the monomer concentration, while at the maximum, when $i = 60$, aggregates of this size are present at about the same concentration as monomer. So in the range of $\ln x_1$ from -8.05 to -8.00 the mole fraction of micelles with an aggregation number of about 60 rises from about $0.05x_1$ to x_1. The micellisation process thus sets in, according to this picture, when x_1 rises from 3.19×10^{-4} to 3.39×10^{-4}.* This is brought out even more clearly in Figure 11.6, where the fraction of the added monomer which is present as micelles is shown as a function of i for various values of $\ln x_1$. The values of the c.m.c. and the most probable micelle size depend on the choice of a, b, and c in equation (11.6); the mean size is determined by b and c, while for given values of b and c the c.m.c. decreases as a decreases.

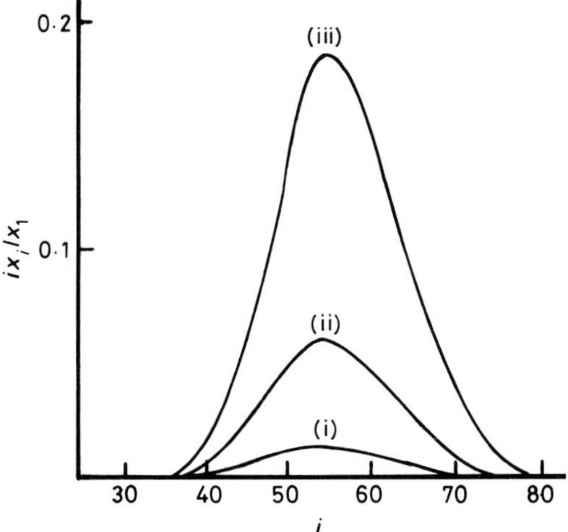

Figure 11.6 *Ratio of the fraction of material in micelles of size i to the mole fraction of monomer as a function of i, for various monomer concentrations. Curve* (i): $\ln x_1 = -8.15$; *curve* (ii): $\ln x_1 = -8.12$; *curve* (iii): $\ln x_1 = -8.10$.

* The c.m.c. is strictly given by the total concentration of the surfactant $x(\text{total}) = ix_i$. This has to be taken into account in any more detailed theory, but the above arguments give a reasonable picture of the sharpness of the onset of micellisation.

This discussion illustrates the main features of the micellisation phenomenon. It remains to say a few words about the significance of the various terms in equation (11.6). The negative contribution from the first term on the right arises from the free-energy decrease when hydrocarbon/water interfaces are replaced by hydrocarbon/hydrocarbon interactions. The second term may be interpreted as a contribution from the excess free energy of the surface of the micelle, while the third arises from the increasing head-group interactions as i increases.

A more detailed theory examines the origin of the free-energy change on eliminating hydrocarbon/water interactions. There is a good deal of evidence that the main contribution to this decrease arises from the increase in entropy when a monomer molecule is incorporated into a micelle. This is at first sight in conflict with the statement made on page 127. However, it is closely bound up with the structural changes which occur when a hydrocarbon chain is transferred from a water to a hydrocarbon environment. Two effects may contribute. It is generally supposed that water around an inert hydrocarbon chain is, in some rather ill-defined way, more structured than in bulk water; removal of the hydrocarbon chain into a micelle allows the water molecules to relax into a state of higher entropy. At the same time a hydrocarbon chain surrounded by structured water is restricted in its movement; when in the core of a micelle, it has greater freedom and a higher entropy. These effects have a much wider relevance than to micellisation, but it is not possible to pursue this topic in greater detail here.

The structures of successive telomers may be represented schematically by diagrams such as those in Figure 11.7. These suggest that, initially, addition of an extra molecule to the complex only partly eliminates hydrocarbon/water interactions, but, as the complex grows, an increasing proportion of the surface of an added molecule makes contact with other hydrocarbon segments. The maximum degree of hydrocarbon/hydrocarbon interaction is reached when the spherical structure is completed. But when the spherical shape is approached two further factors become important. First, it becomes more difficult to insert an extra hydrocarbon chain into the core, and, secondly, the density of packing of head groups on the surface is such that the repulsion between them becomes increasingly significant. The above qualitative considerations can be put into a more quantita-

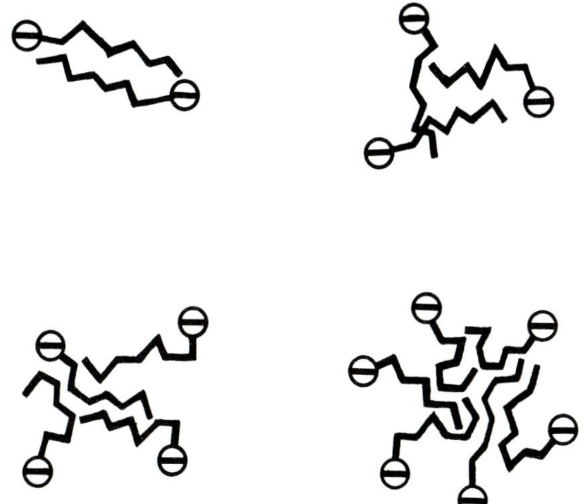

Figure 11.7 *Successive stages in the growth of a micelle. The larger the micelle the greater the loss of water/hydrocarbon contacts and the gain in hydrocarbon/hydrocarbon contacts when a surfactant molecule is added to the micelle.*

tive form and lead to equations such as (11.6).

So far we have considered only the case of a spherical micelle. As more surfactant is added to the solution above the c.m.c., this can be accommodated either by increasing the number of micelles of a given size or by modifying the micellar structure to allow a larger number of molecules to be incorporated into each micelle. This can be achieved by a change of shape of the micelle to a disc-like or cylindrical micelle [Figure 11.2(b) and (c)]. This is possible because the repulsion between head groups can be decreased by this change in structure, and the onset of the repulsive contribution to $\Delta G^{\ominus}(i)$ becomes effective only at larger aggregation numbers. Eventually, a further structural change may occur leading to lamellar micelles [Figure 11.2(d)], reminiscent of biological bilayer films, again allowing head-group interactions to be reduced.

Mention must also be made of yet another structure capable of satisfying the geometrical and energetic criteria. The system may form *vesicles* [Figure 11.2(e)], in which the spherical bilayer shell has an aqueous phase both within and without. This is even more suggestive of a biological cell, and studies of the properties of vesicles represent an important contribution of colloid science to biology.

SOLUBILISATION

The core of a micelle has been shown to have properties akin to a liquid hydrocarbon. One piece of evidence for this is that a surfactant solution above the c.m.c. is capable of taking up substantial quantities of non-polar organic (lipophilic) substances. These enter the core of the micelle, which can now swell because the added material has no hydrophilic moiety which needs to be on the surface (Figure 11.8). *Solubilisation* plays an important role in detergency and in the detailed mechanism of emulsion polymerisation.

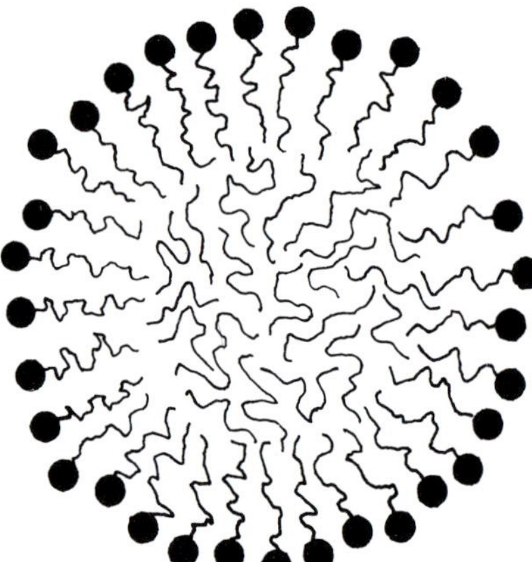

Figure 11.8 *Solubilisation involves the incorporation of non-polar organic material into the core of a micelle.*

Chapter 12

Thin Films, Foams, and Emulsions

INTRODUCTION

So far we have been concerned mainly with dispersions of solids in which all three dimensions of the particles of the dispersed phase are in the colloid size range. An important group of colloidal phenomena involve systems having only one dimension in this range (see Figure 1.1). This chapter deals with thin liquid films, present in isolation, as the basic components of liquid foams or as the film between two emulsion droplets in contact. Other examples of such colloidal systems are those in which the solid particles are thin plates, as in many clay systems, and solid foams. These will, however, be omitted from the present discussion.

SURFACE TENSION AND FILM TENSION

We consider the simple example of a soap film supported on a rectangular frame and held in equilibrium by a force exerted on a moveable edge (Figure 12.1). The work done in stretching the film, and increasing its area by dA, is

$$\mathrm{d}W = -\mathcal{F}\mathrm{d}x = -(\mathcal{F}/l)l\mathrm{d}x = -(\mathcal{F}/l)\mathrm{d}A = -\Sigma_{\mathrm{f}}\mathrm{d}A. \quad (12.1)$$

The force exerted by unit length of the film is called the *film tension*, Σ_{f}. Since the film consists of two parallel liquid/gas interfaces, we may attribute the film tension to the summation of the surface tensions of the two surfaces:

$$\Sigma_{\mathrm{f}} = 2\sigma^{l\mathrm{v}}. \quad (12.2)$$

However, σ^{lv} is not in general equal to the surface tension of the bulk soap solution.

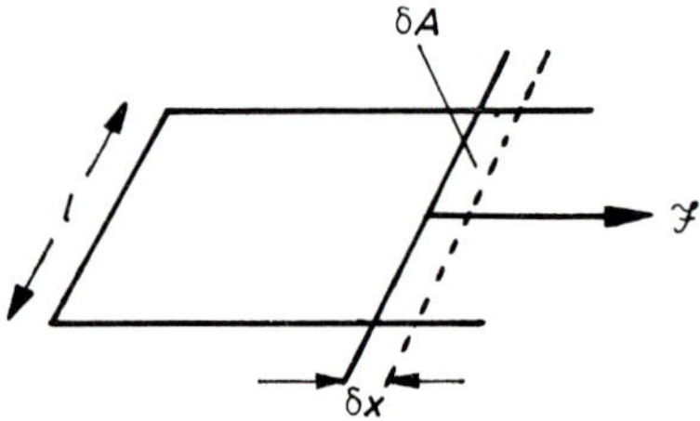

Figure 12.1 *Soap film on a rectangular frame with one moveable side held in equilibrium by a force* $\mathcal{F}$: $\mathcal{F}/l = \Sigma_f$.

SOAP FILMS AND SOAP BUBBLES

The study of the shapes of soap films and bubbles is a fascinating subject. For an account of many simple experiments the reader is referred to the classical book (1890) by C. V. Boys, 'Soap Bubbles and the Forces which Mould Them', reprinted in the Science Study Series by Heinemann, London, 1960, and by Dover, New York, 1959, or to the more recent book by C. Isenberg, 'Science of Soap Films and Soap Bubbles', Tieto Ltd., Clevedon, 1978.

Here we shall only consider some of the basic principles. We have already seen (Chapter 5, page 73) that the pressure difference across a curved surface separating two phases α and β (Figure 12.2) is given by the Laplace equation. In the case of a soap film this is

$$\Delta p = p^{\beta} - p^{\alpha} = C\Sigma_{\mathrm{f}}, \tag{12.3}$$

where the film tension replaces the surface tension. As before, the curvature is defined as

$$C = \frac{1}{r_1} + \frac{1}{r_2}, \tag{12.4}$$

where r_1 and r_2 are the principal radii of curvature of the surface. As noted previously, r_1 is given a positive sign if the centre of curvature of r_1 lies in phase β, and similarly for r_2. These radii may be of opposite signs, as when the surface has a saddle shape (Figure 12.3). In particular, if $r_1 = -r_2$, the pressure difference is zero; this is the case, for example, when a soap film is formed between two parallel circular rings (Figure 12.4), provided that the separation between the rings is less than 1.325 times the radius of the rings.* This is but one of a large variety of shapes which soap films can adopt depending on the geometry of the constraining wires or walls. Equation (12.3) shows, however, that provided the air pressures inside and outside the bubble are constant (but not necessarily equal) the surfaces of bubbles must be *surfaces of uniform curvature*.

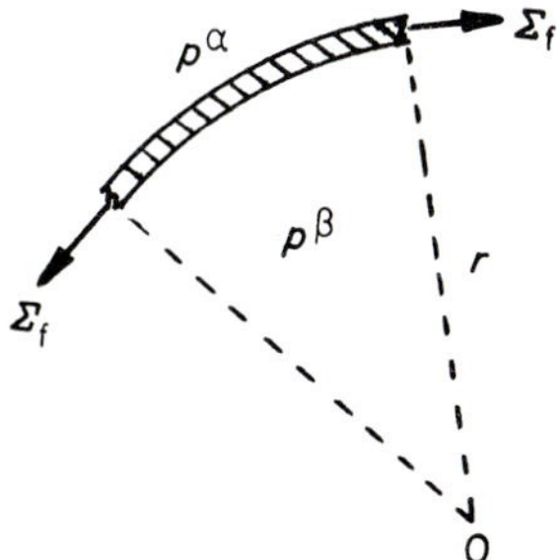

Figure 12.2 *Pressure difference $\Delta p = p^\beta - p^\alpha$ across a film between phases α and β exerting a film tension Σ_f.*

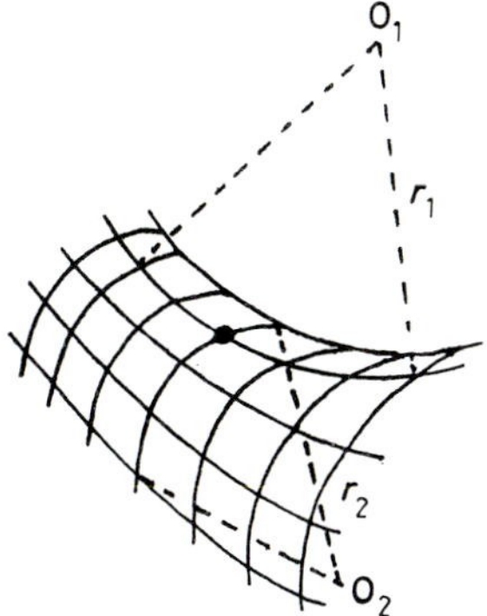

Figure 12.3 *Surface with principal radii of curvature of opposite signs.*

* At greater separations the film becomes mechanically unstable and bursts, to form two planar films filling the wire circles.

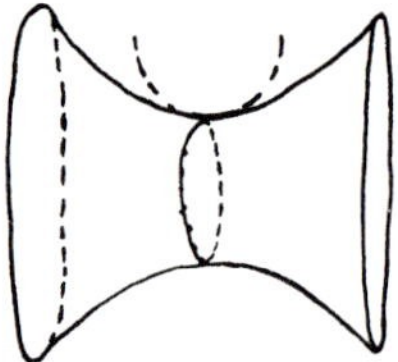

Figure 12.4 *Soap film between two rings: the principal radii of curvature are equal and opposite. The surface has zero curvature ($C = 0$), and there is no pressure difference across the surfaces.*

FILM STABILITY

In discussing film stability it is instructive to follow the same procedure as we did for particulate dispersions and to consider in turn the factors tending to thin the film and those which prevent its collapse and so stabilise it at an equilibrium thickness.

In Chapter 3 we saw that the energy of van der Waals interaction between two planar surfaces of material 1, *e.g.* a liquid or a solid, separated by an intervening medium 2, *e.g.* air, of thickness H [Figure 12.5(a)] is given by equation (3.11):

$$\Delta G^{\mathrm{att}} = -A_{\mathrm{H}}/(12\pi H^2), \tag{12.5}$$

where A_{H} is a composite Hamaker constant given by [equation (3.14)]

$$A_{\mathrm{H}} = [A_{10}^{1/2} - A_{20}^{1/2}]^2. \tag{12.6}$$

Since equation (12.6) is symmetrical in 1 and 2, the same equation applies (perhaps surprisingly at first) to the free energy of a film of 1 in air [Figure 12.5(b)]. This free energy becomes increasingly negative as H decreases (Figure 12.6) so that van der Waals forces, if not opposed by repulsive forces between the surfaces, will cause the film to thin and eventually burst. Repulsive forces are absent in the case of pure liquids, so that they do not form stable films.*

To understand the stability of films we must, just as in the case of particulate dispersions, seek sources of a repulsive potential which resists the thinning process. Such repulsive forces develop when the liquid is a solution of a surfactant which is adsorbed at the liquid/vapour interfaces; they can arise from electrostatic,

* Thin films of liquid metals may constitute an exception to this generalisation.

steric, or adsorption effects, although other specific effects sometimes contribute.

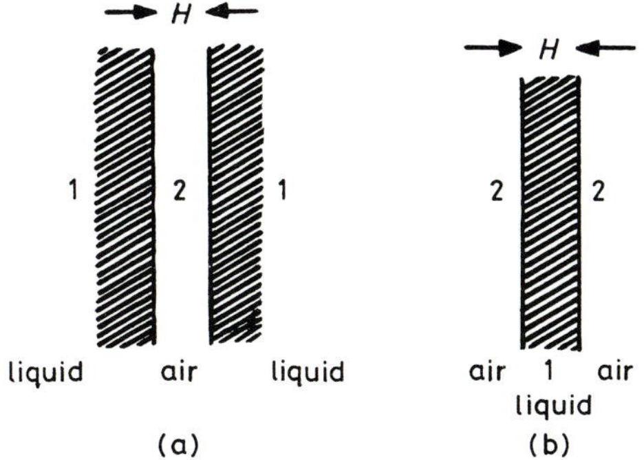

Figure 12.5 (a) *Layer of material* 2 *(e.g. air) between two solid or liquid surfaces of material* 1, *separated by a distance* H; (b) *film of liquid* 1 *between two regions of* 2 *(air).*

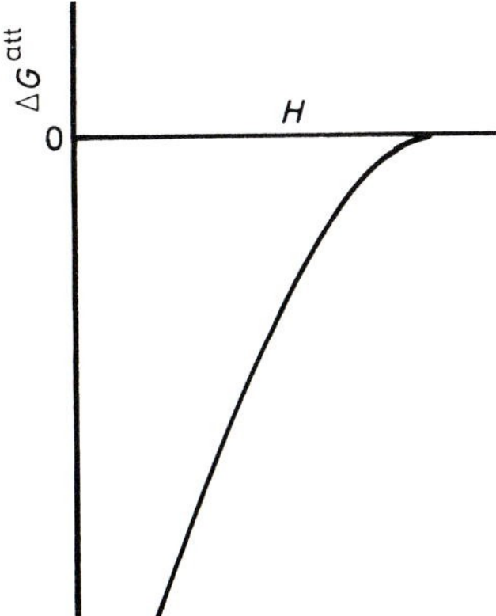

Figure 12.6 *Free energy of a film as a function of its thickness* (H) *arising from van der Waals forces.*

In aqueous solutions of ionic surfactants, electrostatic forces usually provide the dominant contribution. Thus, when an ionic surfactant is adsorbed at a liquid/air interface, it establishes an electric charge at the interface which must be balanced by a diffuse double layer extending into the bulk liquid [Figure 12.7(a) and (b)]. If two such surfaces come together 'back-to-back', as in a thinning liquid film, the two double layers repel one another and a repulsive potential results (Figure 12.8). As before (Chapter 3), the range and magnitude of this repulsion decrease with increase in ionic strength. The overall result is that, just as in the case of interacting particles, the total free energy, which is the sum of the interaction and electrostatic potentials, may take on a variety of forms which determine the stability or otherwise of the film.

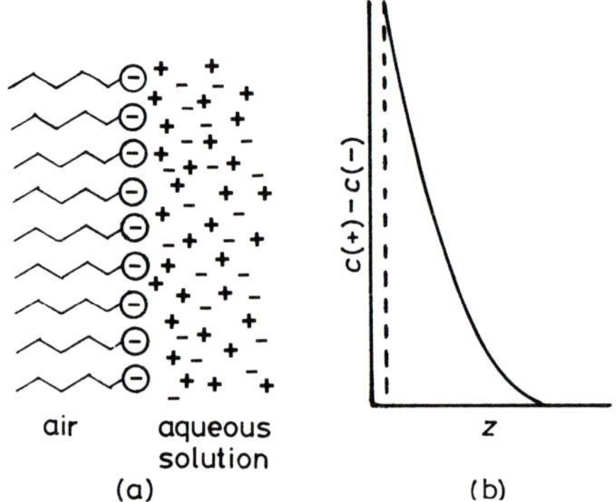

Figure 12.7 (a) *Electrical double layer created by adsorption of an anionic surfactant at the liquid/air interface;* (b) *excess positive charge in the diffuse double layer as a function of distance from the surface.*

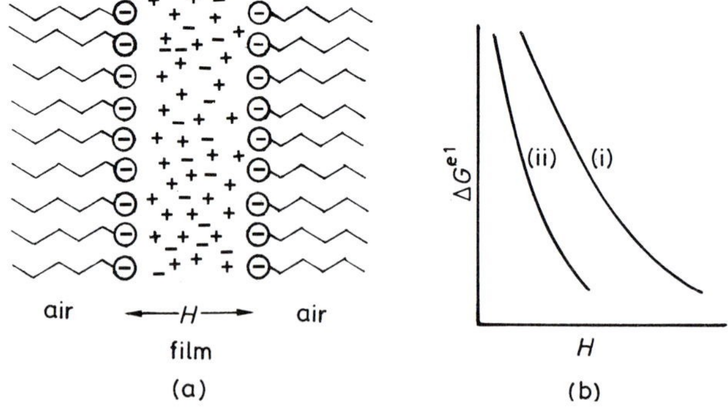

Figure 12.8 (a) *Approach of two charged surfaces of a film 'back-to-back';* (b) *free energy of interaction as a function of film thickness (H), for two charged surfaces. Curves* (i) *and* (ii) *refer to lower and higher electrolyte concentrations respectively.*

On the other hand, certain non-ionic surfactants, especially polymeric surfactants, can exhibit steric repulsion. This arises when the polymeric chain is hydrophilic and is attached to a hydrophobic head group (Figure 12.9). The film-forming properties of such surfactants will be expected to be roughly related to their HLB (Chapter 11).

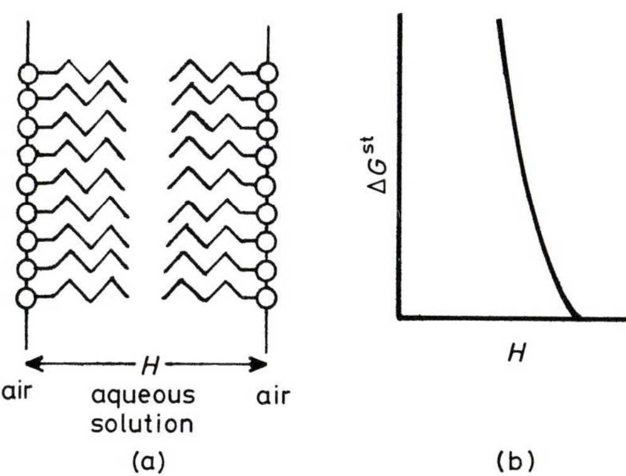

Figure 12.9 (a) *A film stabilised by a non-ionic surfactant;* (b) *free energy of interaction as a function of film thickness for a sterically stabilised film.*

In most examples of aqueous films between vapour phases, the total free-energy curves are believed to have one of the forms shown in Figure 12.10.

If the double-layer repulsion is sufficiently large, the free-energy curve may have the form of Figure 12.10(a). The film is a stable, thick film which shows interference colours indicating thicknesses in the range 300—500 nm. A curve such as that in Figure 12.10(b) leads to a stable film of thickness H_f, which decreases as the electrolyte concentration increases and the range of the double-layer repulsion becomes shorter. A third type of curve [Figure 12.10(c)] has two minima – a primary minimum (pm) and a secondary minimum (sm). As the film drains, it forms a metastable film of thickness H_{sm}. If the energy barrier is not too high, it may convert to a thinner stable film of thickness H_{pm}. It turns out that the thickness of these films is such (5—50 nm) that they do not show interference colours but appear black. The thicker film is called the *first* (or *common*) *black film*, while the thinner is the *second* (or *Newton*) *black film*.* Aqueous soap films exhibit this phenomenon.

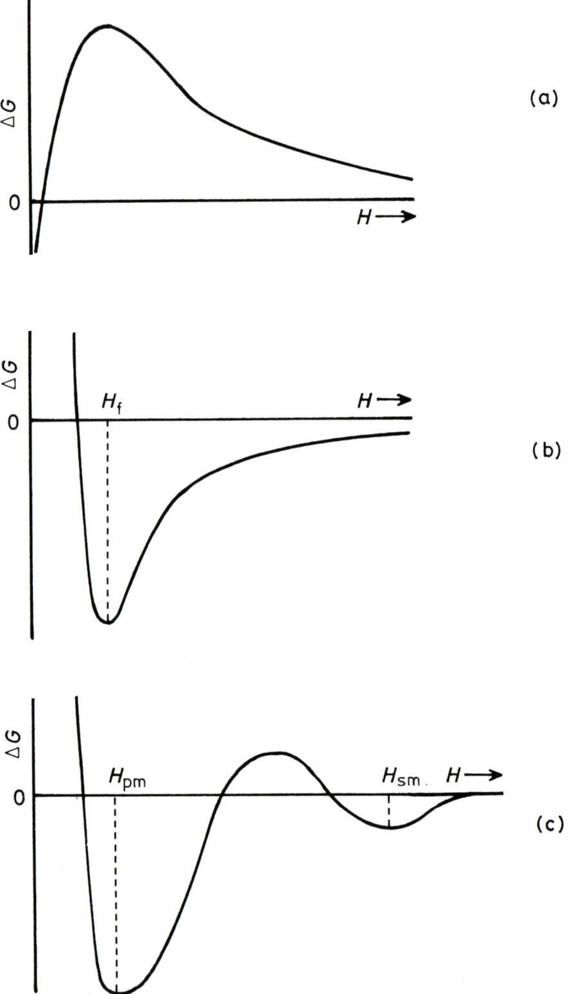

Figure 12.10 (a) *Free-energy curve leading to a thick film;* (b) *free-energy curve leading to a stable film;* (c) *free-energy curve leading to common and Newton black films.*

* It was Isaac Newton who, following Hooke's observation of 'holes' in a soap bubble just before it bursts, described so-called 'black' films of various shades.

The transitions from an unstable thick film to black films are easily demonstrated by allowing a vertical soap film supported on a frame to drain (Figure 12.11). Initially the whole film shows interference colours, then a dark boundary, separated from the

Figure 12.11 *Draining soap film showing interference colours from thicker films and silver and black bands from thin films. The two types of black film are not distinguishable.*
(Photograph: D. Jones, School of Chemistry, University of Bristol)

coloured area by a silvery band, appears at the top and moves downward. This then divides into two areas occupied by the common black and Newton black films. The boundary between the two is not easily visible but can be detected from the differing extents to which they reflect light. Measurements of this kind show that common black films are about 30 nm thick and Newton black films are around 5 nm thick. This latter distance corresponds to the thickness of a bilayer of soap molecules (Figure 12.12).

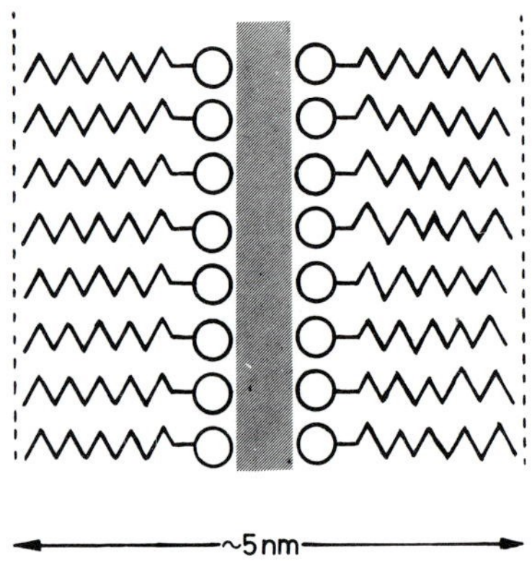

Figure 12.12 *Schematic representation of the structure of a Newton black film. The core (shaded) contains water and counter-ions.*

FILM ELASTICITY

Stable films are characterised by the property that a fluctuation in thickness leads to an increase in free energy, as indicated by the above diagrams (*cf.* Chapter 2). These refer, however, to changes that occur slowly so that equilibrium is maintained. However, many of the disturbances to which films are subjected (*e.g.* vibration, draughts, *etc.*) are of short duration. Under these conditions another mechanism, proposed by Gibbs and Marangoni, operates. If an element of film is stretched quickly, the local surface concentration of surfactant is decreased and, in accordance with the Gibbs equation (5.2), the film tension increases, thus opposing stretching. Similarly, on compression the surface concentration increases, the local film tension decreases, and the compressed region is re-stretched by the surface tension gradient.* Thus the film exhibits elasticity. However, if the fluctuation is too large, the surface concentration may be so reduced that the adsorbed molecules no longer present a sufficient barrier to film thinning and the film will collapse. A hole once formed propagates spontaneously, reducing the surface area of the system, and the whole film is disrupted.

FOAMS

Foams are produced when gas is blown into a liquid through one or more small orifices or through a sintered glass disc, or in many cases when the liquid and gas are simply shaken together. Foams consist of relatively coarse dispersions of gas bubbles in the liquid, in which the gas volume is very much greater than that of the liquid. In most foams the gas bubbles, whose sizes are not normally in the colloid size range, are initially spherical but rapidly adopt a close-packed structure in which the individual bubbles are separated by a thin liquid film of colloidal thickness. The pressure difference across the interfaces is usually very small, so that the individual cells of the foam consist of polyhedra bounded by plane faces (Figure 12.13). The forces that maintain equilibrium along the lines where two faces meet, and at corners, require more detailed consideration which we do not have space here to examine.

Individual isolated films may be stable over long periods, especially when in contact with saturated vapour so that evaporation is prevented. Foams are generally much less stable. This is because, under the influence of gravity, liquid drains from them along the channels (*Plateau borders*) formed at the junctions between film lamellae (Figure 12.14). The gravitational forces are enhanced by surface tension forces: it is seen that in the neighbourhood of the junction there are sharply curved surfaces. The resultant reduced pressure in these regions sucks material from the film and may overcome the forces that stabilise the film,

* Processes that occur as a result of gradients of surface tension are called *Marangoni effects*. A commonly quoted example is the formation of 'tears of wine' on the side of a glass containing wine of sufficient strength. Preferential evaporation of alcohol from the film wetting the glass causes the surface tension of the film to increase and suck liquid up the side of the glass. The liquid then forms tears that run back into the glass.

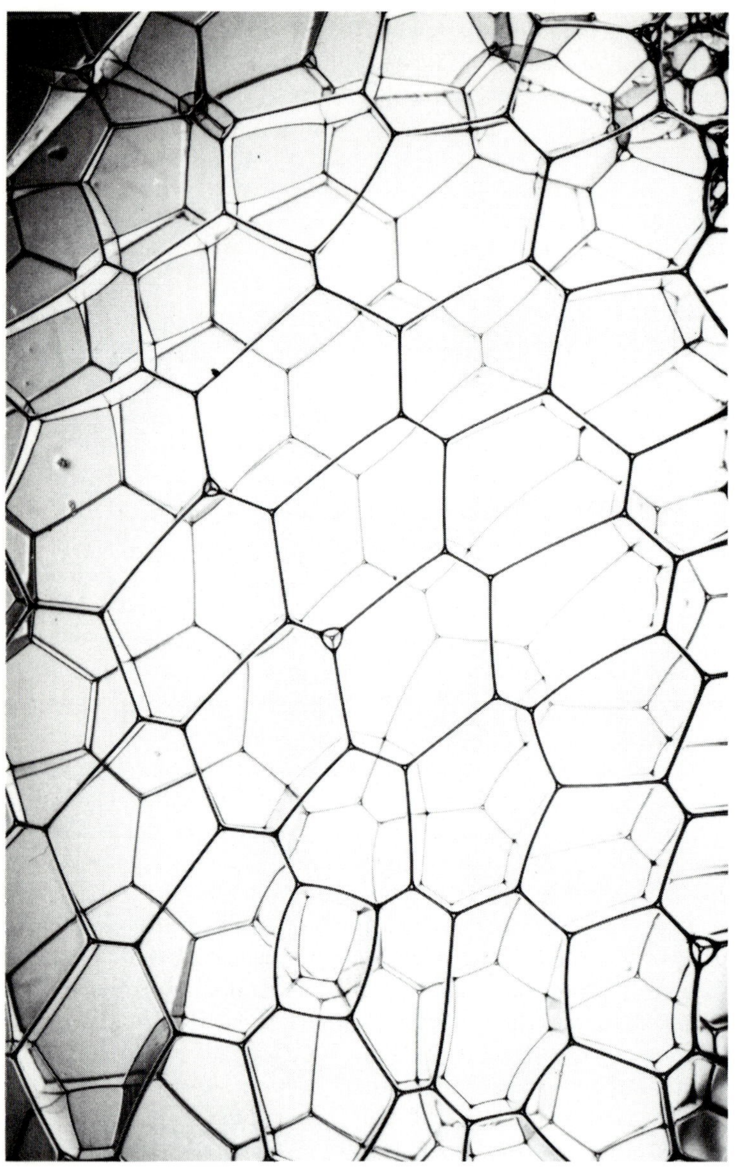

Figure 12.13 *Photograph of a foam showing the polyhedral cells bounded by plane faces.*
(Photograph: D. Jones, School of Chemistry, University of Bristol)

leading to its collapse. Among other factors influencing the lifetime of a foam is the viscosity of the solution, which determines the rate at which drainage occurs.

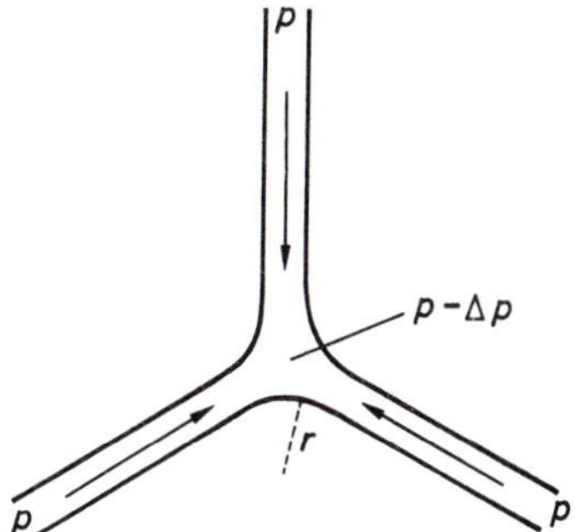

Figure 12.14 *Plateau border at the junction of film lamellae. The channel down the centre provides a route for drainage of the foam. The pressure drop Δp sucks liquid from the foam lamellae.*

FOAMING AND ANTIFOAMING AGENTS

One important group of foaming agents are the fatty-acid soaps and synthetic surfactants. They are generally most effective at or around the c.m.c., when, at the liquid/vapour interface, they form compact monolayers. Frequently their effectiveness is enhanced by the addition of a second component. Thus a mixture of lauryl (dodecyl) sulphate and lauryl (dodecyl) alcohol forms a much more stable foam than the sulphate alone. Non-ionic surfactants with an appropriate HLB are also good foaming agents; these include various water-soluble polymers. Although the basic theoretical ideas discussed above give useful clues as to the effectiveness of foaming agents, in practice their formulation is still largely an empirical procedure.

An important group of foaming agents which differ in their mode of action from surfactants are certain proteins. In this case soluble proteins are irreversibly 'denatured' on adsorption and agitation. They become insoluble and form multi-molecular, rigid, surface layers which produce very long-lasting foams. Indeed many of the foams familiar in the kitchen are of this type: examples are beaten egg-white, whipped cream, and mousse.

Yet another mode of foam stabilisation is by finely divided, hydrophobic solid particles such as coal-dust or particles of metal soaps. Their effectiveness arises from their aggregation at the film surfaces: a closely packed layer of solid particles prevents coalescence. However, the phenomenon is very dependent on the surface properties of the solid, since it is necessary for the contact angles to be such that the particles are held in the film. The details of the mechanism have not been fully established, but Figure 12.15(a) suggests a possible configuration leading to stabilisation.

It is often necessary to destroy unwanted foams. In some cases this may be done mechanically, for example by blowing a stream of hot gas across the surface of the foam. More often other methods employing *antifoaming agents* are used. These may operate in several ways, not all of which are fully understood. If an agent, not itself able to stabilise a foam, is adsorbed at the surface more strongly than the foaming agent, then desorption, or sweeping away, of the surfactant leads to collapse of the foam. Thus a drop of ether, n-butanol, or capric acid added to a cylinder of foam causes its instant collapse. Many antifoaming agents are insoluble in water and are applied as a dispersion of fine dropets; silicone oils are particularly effective. Under certain circumstances strongly hydrophobic solid particles have antifoaming properties. Thus it is observed that a foam bath is rapidly destroyed as soon as washing is commenced: this happens if the water is hard because the calcium soap particles act as antifoaming agents. Finely divided silica also has antifoaming properties. Whether or not hydrophobic solid particles act as foam stabilisers or destroyers depends both on a subtle balance of surface properties and on particle size. Antifoaming action by solid particles is thought to result from the bridging by the solid between the two film surfaces, a process enhanced when the contact angle at the air/film/solid interface approaches 180° [Figure 12.15(b)].

FROTH FLOTATION

The process of *froth flotation* for the separation of minerals exploits the fact that either minerals differ naturally in their degree of wettability by surfactant solutions or their surface properties can be modified by adsorption. When air is blown through a suspension of finely ground ore in an appropriate surfactant solution, those particles which have the relevant surface properties become attached to the bubbles and rise to the surface. The froth containing these particles can then be swept away and collected as the foam breaks. For a particle to be collected it must, on collision with a bubble, become attached to it. The condition for this is that the contact angle should be greater than about 50°, when the surface/particle configuration shown in Figure 12.16 is

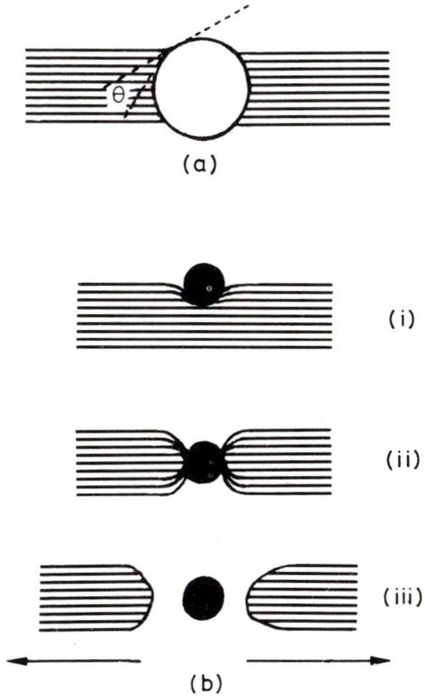

Figure 12.15 (a) *Possible configuration of a small solid particle acting as a foam stabiliser,* $\theta < 90°$; (b) *strongly hydrophobic particle acting as a foam breaker:* (i) *particle making contact with one side of the film,* (ii) *particle with a contact angle of 180° bridging the film,* (iii) *expansion of the hole created by the particle.*

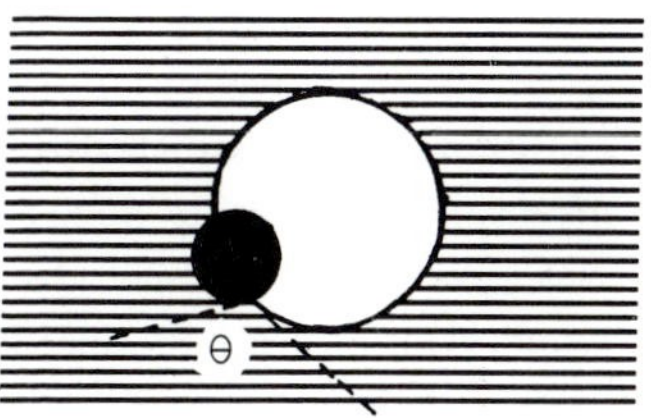

Figure 12.16 *Attachment of a solid particle to a bubble.*

established. The choice of a suitable mixture of surfactants for a particular mineral process is still largely empirical, but it nearly always involves the use of both a surfactant which is adsorbed at the mineral surface to render it hydrophobic (a *collector*) and one which produces a foam (*foamer*).

EMULSIONS AND MICROEMULSIONS

In emulsions the dispersed phase and the dispersion medium are both fluids. The commonest examples are those in which the two phases are oil and an aqueous medium. They may be of two distinct types: a dispersion of fine oil droplets in an aqueous medium, an *oil-in-water* (O/W) emulsion, or of aqueous droplets in oil, a *water-in-oil* (W/O) emulsion. In some special cases a *bicontinuous* emulsion may be formed in which one phase forms a continuous network in the other. Recently, dilute *gas-in-liquid* emulsions (dispersions of fine gas bubbles in liquid) have been shown to exist in solutions of gas at high pressure in liquids (for example in carbonated drinks). From this point of view also an aerosol of liquid droplets may be regarded as a dilute liquid-in-gas emulsion.

Whether or not an O/W or W/O emulsion is formed depends on a number of factors. If the ratio of the amounts of the two phases (the ratio of *phase volumes*) is very low, the phase present in the lower amount is often the disperse phase; if the phase volumes are roughly equal, other factors determine which type of emulsion is formed. It is usually possible to identify the type of emulsion by examining the effect of diluting it with one of the phases. Thus on adding water to an O/W emulsion the emulsion is diluted. On the other hand, on adding oil the added oil forms a separate layer. That milk may be diluted with water shows that it is an O/W emulsion, while mayonnaise, a W/O emulsion, can be blended with additional oil.

As already discussed in earlier chapters, the formation of an emulsion, which involves an increase in interfacial area between the two phases, is accompanied by an increase in free energy. The ease of formation of an emulsion may be measured by the amount of work needed for its formation: the lower the interfacial tension the less work is needed and the more readily the emulsion is formed. The addition of *emulsifying agents* which adsorb at the interface and lower the interfacial tension is therefore usually necessary if a stable emulsion is to be formed. Very few emulsions exist in which both phases are rigorously pure liquids. One will expect emulsifying agents for O/W or W/O emulsions to be amphipathic and to have an appropriate HLB, so that adsorption at the interface involves the orientation of the adsorbed molecules with the hydrophilic moiety lying in the aqeuous phase and the lipophilic portion in the oil phase.

Emulsions, like foams, can also be stabilised by finely divided solids, provided the properties of the solid/liquid/liquid interface are appropriately adjusted. These properties may also determine whether an O/W or a W/O emulsion is formed. For example, shaking water and benzene together with finely divided calcium carbonate yields a benzene-in-water emulsion. On the other hand, if the calcium carbonate is made hydrophobic by treatment with oleic acid solution, a water-in-benzene emulsion results.

Although lowering of the interfacial tension is a prerequisite for emulsion formation, it is not by any means the only factor controlling stability. The process of de-emulsification involves a series of steps. Thus the droplets have first to come into near contact. Factors such as droplet charge and the presence of a steric stabilising layer may lead to a repulsion between droplets, tending to maintain their stability. In addition, as in the case of solid particles in a liquid, the approach of two droplets involves the expulsion of continuous phase from between them, and this will be influenced by the rheological properties of the continuous phase. The added complication here is that the surfaces of the droplets are not rigid, and as mentioned previously (Figures 10.4 and 10.6) they deform on collision. Finally, in contrast to the case of a dispersion of solid particles, the breaking of an emulsion involves the loss of the individuality of the droplets which form a larger droplet of lower surface/volume ratio. In this process the adsorbed layers have to be displaced and the final step is controlled by the stability of the thin film between them. The considerations discussed on page 173 then become important, and it is necessary to understand the properties of thin films before emulsion stability can be fully explained. In particular, highly viscous or visco-elastic films tend to confer stability on an emulsion.

De-emulsification is important, for example in the oil industry. Agents for this purpose have to be capable of displacing the stabilising surfactant, leaving a non-stabilising interface. The precise nature of those in use is not revealed, but many of them are polymeric materials.

The ease of emulsion formation increases and the droplet size achievable decreases as the interfacial tension falls. Systems in which the interfacial tension falls to near zero [$<10^{-3}$ mN m^{-1} (dyne cm^{-1})] may emulsify spontaneously under the influence of thermal energy and produce droplets so small (<10 nm diameter) that they scatter little light and give rise to clear dispersions. The *microemulsions* so formed occupy a place between coarse emulsions and micelles. They are usually effectively monodisperse and unlike coarse emulsions are thermodynamically stable. Microemulsion droplets have sometimes been classified as swollen micelles. In fact, there probably exists an essentially continuous sequence of states from association colloids to coarse emulsions.

Microemulsions were first discovered empirically by Schulman, who found that the addition of a fourth component (often an alcohol) to an emulsion containing oil, water, and a surfactant led to the formation of a clear, apparently homogeneous phase. This additional component is usually called the *co-surfactant*. Microemulsions have been the subject of intense study in recent years, especially in view of their possible use in enhanced oil recovery. A typical recipe for forming a microemulsion is given in Appendix I.

Chapter 13

Gels

INTRODUCTION

Gels are essentially dispersions in which the attractive interactions between the elements of the disperse phase are so strong that the whole system develops a rigid network structure and, under small stresses, behaves elastically. In some instances the gel may flow plastically above a limiting yield stress [Chapter 8, Figure 8.3(e)]; the gel then often exhibits thixotropic behaviour, the rigid gel state being re-formed when the stress is removed.

The disperse phase may consist of solid particles (*e.g.* clay platelets), macromolecules (*e.g.* gelatine), or surfactant molecules (*e.g.* soaps). Thus concentrated suspensions of clays or latex particles form gels; some examples are certain densely filled lubricating greases (inorganic gel greases) and gelled paints. Familiar domestic examples of macromolecular gels are those formed by hydrophilic polymers (mostly of biological origin) which swell spontaneously in contact with water but, because of the strong bonds between individual molecules, retain a degree of rigidity. In certain concentration ranges soap solutions also form gels. Rubber swollen by aromatic hydrocarbons is another example.

FORCES LEADING TO GEL FORMATION

Several types of interaction may lead to gel formation. These may be electrostatic, van der Waals interactions, or chemical bonding.

In the case of charge-stabilised dispersions above a certain concentration, addition of electrolyte often leads to gelation rather than the formation of distinct flocs. Thus gelation may be regarded as the formation of an extended, continuous floc filling the whole system (*cf.* Figure 9.8). In general, more or less spherical particles tend to associate as 'strings of beads', while rod-like particles may build up a 'scaffolding' or 'brush-heap' structure, as in the case of vanadium pentoxide gels. Clay platelets, such as those of montmorillonite, form a structure, often described as a 'house-of-cards', controlled largely by electrostatic forces. This is particularly marked in the case of kaolinite, the platelets of which carry a negative charge on the flat surface of the crystallite but, at low pH, a positive charge around the edges [Figure 13.1; *cf.* Figure 3.3(e) and (f)]. As a result, edge-to-face attractions stabilise the structure (Figure 13.2). This very open network means that a gel can be formed in quite dilute aqueous dispersions (around 2%). If this structure is broken down, by for example the application of pressure, the platelets adopt a parallel arrangement of greater density whose properties are controlled by the electrostatic repulsion between parallel charged surfaces. Similarly, if the pH is raised, the positive charge on the edges decreases, the edge-to-face interactions weaken, and the house-of-cards collapses to form, again, a parallel configuration. These changes have a major influence on the rheological properties of the gel.

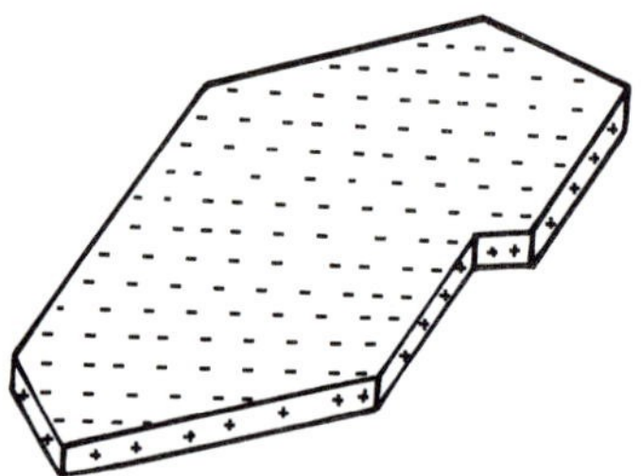

Figure 13.1 *Kaolinite crystal with negatively charged faces and positively charged edges.*

Lamellar gels are also formed by soaps under appropriate conditions of concentration and temperature. These may be regarded as essentially micellar systems [Figure 11.2(d)] in which at high concentrations the lamellar micelles extend over considerable distances and interlock to form a continuous gel network.

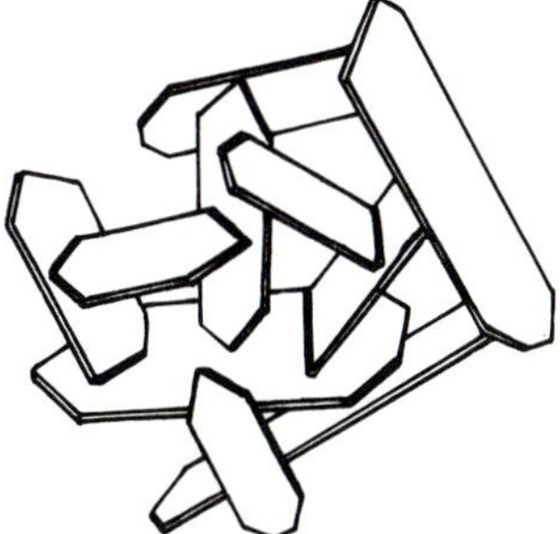

Figure 13.2 *Edge-to-face interaction leading to a 'house-of-cards' structure.*

In macromolecular aqueous gels the network structure is controlled to a major extent by hydrogen bonding, although where the molecules concerned are polyelectrolytes ionic forces also contribute. The polymers that form gels are usually copolymers – often block copolymers – containing polysaccharide and/or protein chains. In dilute solution, and especially at higher temperatures, the polymer exists as a loosely coiled chain [Figure 13.3(a)]. At higher concentrations and lower temperatures portions of the polymer chains, associated with one or other of the polymer blocks, aggregate to form double or triple helices [Figure 13.3(b)], leading to the formation of a network structure. If the polymer segments which aggregate are polyelectrolytes (usually because of the presence of weak carboxylate groups), additional binding may be brought about by the addition of cations, especially divalent cations. For example, in the gelation of alginates (copolymers of mannuronic and guluronic acids) it is necessary to have calcium ions present [Figure 13.3(c)].

Since gel formation involves a balance between the attractive forces tending to form a rigid network and the thermal motion of the polymer chains, gelation processes are markedly dependent both on factors such as pH and electrolyte concentration, which affect the former, and on temperature, which affects the latter. Thus it is commonly observed that a gelatine solution, which is liquid-like at higher temperatures, forms a jelly when cooled. Similarly, if electrolyte is added to a polyelectrolyte solution, leading to increased interchain interactions, gelling may be induced. Gelling in non-aqueous systems may also be produced by adding a non-solvent to an initially stable sol.

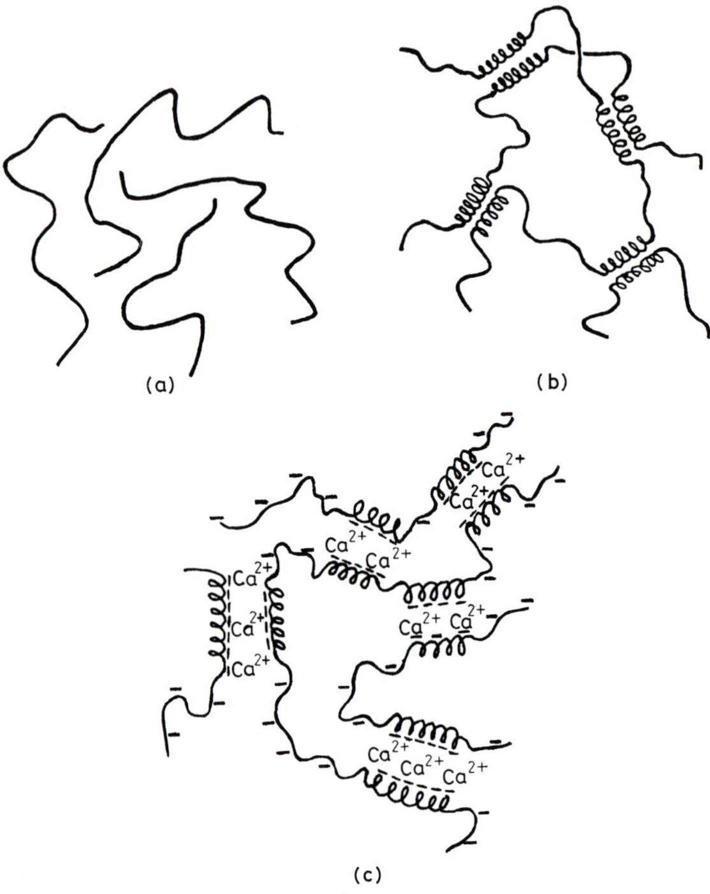

Figure 13.3 *Gelation of polymer solutions:* (a) *polymer chains in free random motion,* (b) *polymer chains forming helices which interact to form a gel structure,* (c) *gel formation in polyelectrolytes, often enhanced by the addition of, e.g.,* Ca^{2+} *ions.*

Processes in which by various procedures a sol is converted into a gel are *sol–gel processes* and have found important industrial applications (see Chapter 14).

Another important class of gels are the rigid gels formed by chemical crosslinking of polymeric systems. To produce these a polymer is reacted with an appropriate crosslinking agent which forms covalent bonds between segments of the polymer. If all segments are equally reactive, the formation of crosslinks will be a random process. Examples or rigid gels of this type are vulcanised rubber and ion exchanger beads. Gels of this kind usually remain elastic up to high stresses.

Many gels can be dried, leaving a porous solid that has a network structure. These are called *xerogels*. Silica gel is a typical example.

SWELLING PROPERTIES OF GELS

Gels often form spontaneously. In the case of aqueous organic gels the macromolecules may be looked upon as containing hydrophilic groups (*e.g.* glucose units) which, in the absence of the network structure, would tend to dissolve in water. The restrictions imposed by the polymer network mean that, although water can diffuse into the gel, the 'soluble' units cannot diffuse out. In effect the network behaves as its own semi-permeable membrane, so that the swelling process may be regarded as an osmotic phenomenon. Imbibition will therefore occur spontaneously and lead to the stretching of the network, which in turn imposes a pressure on the imbibed water. If the cohesion of the network is strong enough, swelling will cease when the internal pressure is equal to the osmotic pressure of the 'internal solution'. If the network is weak and a sufficient supply of water is available, it will disintegrate under the internal pressure and the polymer will go into solution. Even in this state, however, aggregates of polymer may still exist and influence the rheological behaviour of the solution.

A similar situation arises in the case of the swelling of rubber. Here the isoprene units of the rubber molecule, when not part of a rubber molecule, are soluble in aromatic hydrocarbons, but when incorporated in a polymer network they are constrained to remain as part of a gel.

The swelling of clays in water represents an example in which the internal osmotic pressure is associated with double-layer repulsion between the clay platelets.

The swelling of a gel can be prevented by subjecting it to a pressure applied by a piston permeable to the solvent. This is called the equilibrium *swelling pressure*, equilibrium being set up between external solvent at atmospheric pressure and solvent confined in the compressed gel. The swelling pressure depends on the solvent content of the gel, and hence on the degree of swelling, and decreases with the extent of uptake of solvent. In the case of a weak gel, swelling passes over to solution, and the swelling pressure (now a true osmotic pressure) falls to zero at infinite dilution. On the other hand, for a strongly crosslinked gel the swelling pressure decreases to zero at a limiting equilibrium degree of swelling.

Swelling pressure can reach very high values. Thus it is possible to break rocks by driving a dry wooden wedge into a fissure and then wetting the wood. The swelling of clays can also lead to serious problems in civil engineering, since under certain conditions swelling pressures sufficient to lift a moderate-sized building can be developed.

It often happens that the network structure initially formed in a gelling process is not the most stable. The relatively slow diffusion of portions of the polymer chains may lead to the formation of more stable and more compact structures which exert an increased swelling pressure on the imbibed liquid, which then exudes from the gel. This slow expulsion of imbibed liquid is called *syneresis*.

Chapter 14

The Industrial Importance of Colloids

INTRODUCTION

In Chapter 1 some indication was given of the widespread occurrence of colloidal systems and of their importance in everyday life. No less importance attaches to the involvement of colloids in industrial processes and products. A comprehensive survey of colloid technology would occupy several volumes, so that all that can be done here is to draw attention to some of the industries in which the application of colloid science has made a major impact in recent decades.

In the previous chapters we have dealt with some of the problems of colloid science by attempting to answer the following scientific questions.

(i) How can colloidal dispersions be formed?
(ii) What factors determine whether a dispersion, once formed, is stable or not?
(iii) Thus, how can dispersions be kept in the dispersed state, or destroyed?
(iv) What special properties are conferred on a system when its structure puts it into the colloid category?

These questions are all relevant to the solution of the main problems of colloid technology, which can be classified broadly in the following way.

(i) The preparation of stable colloidal dispersions with appropriate properties.
(ii) The use of colloidal dispersions as a step in manufacturing processes.

(iii) The utilisation and exploitation of colloidal phenomena.
(iv) The handling of colloids.
(v) The destruction of unwanted colloids.

Table 14.1 summarises the main industries concerned with each of these aspects, although it is important to realise that nearly always a given industrial problem becomes involved with more than one of them. In the following sections a selection of the topics listed in Table 14.1 will be illustrated.

Table 14.1 *Some problems of colloid technology.*

Industrial aspect	*Main industrial applications*
Preparation of stable colloids	Paints, inks, pharmaceutical and cosmetic products, food products, drilling muds, dyestuffs, agricultural chemicals, fire-fighting foams
Use of colloidal dispersions in a manufacturing process	Ceramic casting, cements and plaster, paper coating, magnetic tapes, photographic products, gas adsorbents and catalyst supports, chromatographic adsorbents, membrane production
Utilisation of colloidal phenomena	Detergency, capillary phenomena (wetting of powders, enhanced oil recovery, water/soil relations), flotation of minerals, adsorption of impurities, solvent recovery, electrolytic painting
Handling of colloids	Rheology (pumping of dispersions and slurries, stirring of reactors), caking and flow of powders
Destruction of unwanted colloids	Water purification, fining of wines and beer, sewage disposal, breaking of oil emulsions and foams, dewatering of sludges, dispersal of aerosols and fogs, disposal of radioactive waste

INDUSTRIAL DISPERSIONS

Here we take as main examples the paint, agrochemical, photographic, printing-ink, pharmaceutical, detergent, dyestuffs, and ceramic industries. One of the main operational requirements in several of these is that the dispersion should remain stable over long periods of storage under a variety of ambient conditions. A range of options is available to achieve this end.

In the paint industry the stability of dispersions of titanium dioxide, in the size range 0.2—0.3 μm, may be controlled either by electrostatic or by steric mechanisms, depending on whether the medium is aqueous or oil-based. In water-based paints the surface charge on the particles arises either from ionisation of surface groups (see Chapter 3), when it depends on the pH, or from the adsorption of an anionic polyelectrolyte. Steric stabilisation in oil-based paints is achieved by using copolymers, for example of the 'comb' type. In these a backbone polymeric chain has stabilising chains grafted to it at regular intervals (Figure 14.1). The backbone is chosen to adsorb strongly on to the pigment surface, while the side chains are 'soluble' in the non-polar medium. Complete stability is, however, not desirable, and a weakly flocculated system is preferred for two reasons. First, this avoids the formation of a hard sediment ('*clay*') when the paint is stored for a long time. Secondly, the weak network structure which is established is resistant to gravitational forces but breaks down under the shear stresses imposed when the paint is brushed on to a surface. This is the basic principle underlying the design of thixotropic, 'non-drip' paints. Paints must also form a surface film which in the cured product encapsulates the pigment particles. The degree of flocculation of the pigment particles in the film must also be carefully controlled to achieve the desired optical properties. Thus for a gloss finish the particles must be deflocculated, while the less ordered structure resulting from flocculation (see Chapter 9) leads to a matt appearance. The resin which forms the film is often present as a latex dispersion.

Figure 14.1 *Schematic representation of a stabiliser of the 'comb polymer' type. The backbone is strongly adsorbed by the particle, while the side chains are 'soluble' in the medium.*

While this must remain stable on storage, the latex particles must coalesce to form a smooth surface when applied to a surface and allowed to dry.

A recent development is the use of 'electrodeposition' of paints, which has found wide application in the automobile industry because of the ability to obtain uniform deposition on and in complex geometrical structures. In this, charged particles are caused to move toward the metal surface under the influence of an electric field and to flocculate on meeting the surface.

It is clear from this brief summary that paint technology involves the intimate interaction of a range of colloid phenomena – stabilisation, rheology, emulsions, and electrophoresis – and is a prime example of the way in which control of these properties can be exploited industrially.

Similar problems arise in the case of agrochemicals. Here the active ingredients are often marketed in a concentrated form ready for dilution before use. Storage represents a major problem since it is particularly important to avoid claying, because any sediment which forms must be readily redispersed on dilution and stirring. Again the use of weak flocculation is often the preferred solution. Thickening or structuring the liquid with a soluble polymer may also be employed to reduce the rate of sedimentation.

The biological activity of a pesticide often depends on small particle size, so that if it is appreciably soluble then particle growth by Ostwald ripening (Chapter 10) may become a problem. This may be overcome by choosing a less soluble form of the ingredient or a medium in which it is less soluble, by avoiding the presence of a wide range of particle sizes, or by the addition of a crystal growth inhibitor.

Many insecticides and fungicides are supplied in the form of solutions in an oily medium which when diluted with water undergo spontaneous emulsification under only very mild agitation. Here too the formulation of these products depends on a subtle choice of medium and added surfactants to obtain the desired result.

Other methods of formulating agrochemical products are to form an O/W emulsion in which the disperse phase is a solution of the active ingredient in the oil, to exploit the ability of micelles and microemulsions to solubilise pesticide molecules (Chapter 11), or to employ multiple emulsions in which an aqueous solution of the pesticide is present as an emulsion inside oil droplets, themselves dispersed in an aqueous phase (Chapter 12).

The colloidal properties of soil represent another aspect of the agricultural importance of colloid science. To achieve a well aerated soil with good drainage properties it is necessary for the soil particles to be relatively loosely packed. In particular, heavy clay soils have to be treated to obtain this structure. Again this requires the soil to be weakly flocculated. Many naturally occurring materials such as humic acid may be employed for this purpose, and empirical agricultural practice over the centuries has established procedures by which the fertility of soil can be enhanced without the use of 'artificial' materials. However, in recent years a number of synthetic polymeric surfactants have been introduced as soil conditioners and their usage is likely to increase, especially where the existing natural fertility needs to be improved.

Surface chemical aspects are of great importance in the application of pesticides by spraying. The formation of aerosol sprays of appropriate droplet size, their impact and adhesion to foliage, and the subsequent spreading of the droplets over what are usually waxy hydrophobic surfaces are all problems that require an understanding of basic science for their solution.

In the manufacture of photographic emulsions the 'speed' and 'grain' of a film depend critically on the particle size of the silver halide crystals. Thus, while it is necessary to avoid Ostwald ripening in, for example, agricultural chemicals, the photographic industry has long exploited this phenomenon to obtain the desired properties in its products. More recently, specially developed technology leading to the production of ultrafine ($<40\ \mu m$), highly monodisperse silver halide crystals is an essential step in the manufacture of holographic film. Finally, colour photography is dependent on the adsorption of dyestuff molecules on the silver halide particles to make them sensitive to light of a particular wavelength and thus calls on many aspects of surface chemistry.

The rheological and wetting properties of dispersions are of prime importance in the coating not only of photographic film but also of paper products and magnetic recording tape.

The technology of printing inks has long depended on the empirical application of colloid technology. The use of 'protective' substances such as natural gums dates back several millenia. They operate mainly through steric stabilisation mechanisms, although since they are often polyelectrolytes charge stabilisation may also play a significant role. But the technical demands of modern high-speed printing impose severe requirements on colloidal properties other than that of long-term stability. The transfer of ink to roller, from roller to typeface, and from typeface to paper at high speed is only achieved satisfactorily if the ink has special rheological properties, and it must also be absorbed rapidly into the porous surface structure of the paper and be held there firmly to avoid subsequent smudging. The recent development of ink-jet printing imposes particularly stringent conditions on the colloidal properties of the ink.

Another industry in which stable dispersions and emulsions are essential is the pharmaceutical industry. Methods of dispersing active ingredients in pastes and ointments and the avoidance of exudation of the medium on long storage (syneresis, Chapter 13) have to be developed. The rapid disintegration of tablets in water and the slow release of drugs taken internally represent extreme types of behaviour, and techniques have to be evolved to meet both limiting requirements. Many drugs taken by inhalation are now dispensed from aerosol cans, and here too it is essential that the active material does not sediment or cream; if it does, it must be readily redispersed by shaking. The same applies to the wide range of traditional medicines dispensed as particulate dispersions and carrying the instruction 'Shake Bottle Before Use'.

Special problems arise in the detergent industry. A primary aim is to achieve efficient wetting of the cloth and then, by manipulating the surface chemistry of the textile/soil particle/surfactant solution interfaces, to remove the soil particle from the cloth. It is important that the soil particle should remain in suspension and must not be redeposited elsewhere: a stable colloid must be produced for subsequent removal by rinsing. A further problem results from the presence of calcium and other salts in domestic water. The calcium has to be sequestered or, if precipitated, it must be in a colloidally stable form. This involves, among other things, peptising the calcium soap and controlling the crystallisation of calcium carbonate by appropriate additives. When one bears in mind that domestic detergent formulations are also expected to bleach, impart some fluorescence, and leave the fibres in a soft condition, it is clear that much skill and experience in addition to basic science are needed to ensure that the various additives do not interact to destroy their effectiveness.

The dyestuffs industry has a major involvement in surface and colloid science. Problems arise, for example, in the complete wetting of fabrics, in the grinding and dispersion of vat dyes, and in dye adsorption on textile fibres. Much of the technology was developed empirically over the centuries, but, more recently, increased understanding of the basic science has led to many improvements both in the processes and in the quality of the final products.

In the ceramic industry the rheology of clay suspensions plays a major role. Their colloidal properties are of especial significance in the process of slip-casting, in which a clay slurry is put in contact with a porous plaster mould, which sucks up the water to leave a solid form ready for firing. Here the suction provided by the porous mould in relation to the rate of flow of water through the clay matrix is important. The structure of the clay slurry, as determined by its degree of dispersion, has also to be controlled by suitable additives.

Closely related to this is the general problem in the whole chemical industry of the separation of solids from liquids. The processes of thickening, flocculation, dewatering, and filtration are all intimately controlled by the forces and structures arising in colloidal systems. The improvements brought about by the use of electroseparation processes depend on the exploitation of the electrokinetic effects discussed in Chapter 6.

COLLOIDS IN THE ENERGY INDUSTRIES

Colloids also play a central role in many aspects of the energy industries, especially the oil, coal, and nuclear industries.

The oil industry is concerned with colloid and surface chemistry in at least five major areas. In oil extraction the complex interaction of surface tension forces, wettability, and rheology determines in large measure the efficiency with which oil can be recovered. The fact that it is only rarely possible to extract more than 50% of the oil present in the field means that high rewards would result from a successful and economic technique for increasing the yield. Among the possible colloid-related methods are the following: the use of surfactants to reduce the oil/water interfacial tension and so to reduce the pressure gradients needed to displace the oil [Figure 14.2(a)]; the injection of a surfactant

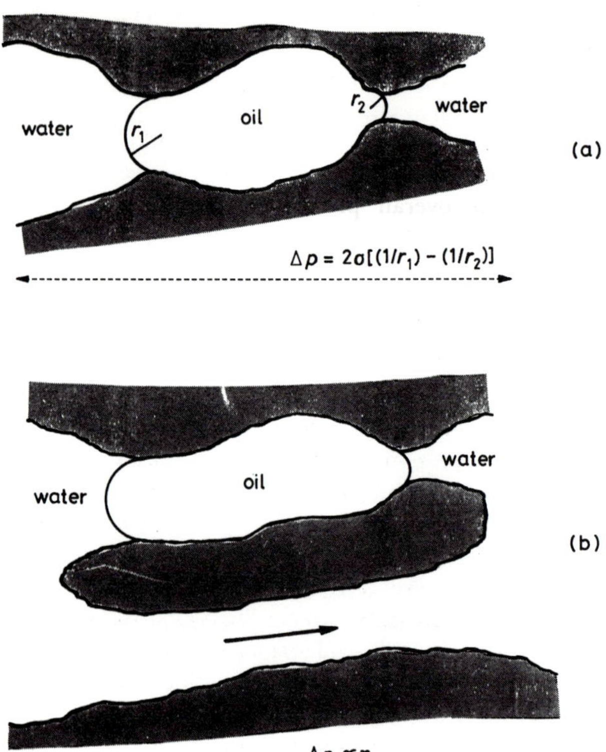

Figure 14.2 *Enhancement of oil recovery.* (a) *To displace an oil blob trapped in a pore, a pressure difference* $\Delta p = 2\sigma[(1/r_1) - (1/r_2)]$ *is needed. This is reduced if the interfacial tension* σ *is lowered.* (b) *If a parallel water channel bypasses the blob, the pressure drop can be increased by raising the viscosity of the aqueous phase by the addition of a polymer.*

'cocktail' capable of causing the oil and water to form a microemulsion which can be swept away; the use of an injected polymer slug to modify the rheology of the process and to inhibit 'fingering' of the water flood through more porous areas of the oilfield [Figure 14.2(b)].

A second area of importance is that concerned with drilling, where 'drilling muds' are injected into the bore hole both to lubricate the drill and to form a coherent lining to the hole. Among the materials currently in use are slurries of derivatives of sodium montmorillonite in oil or water. Surface modification of the clay by ion exchange with a quaternary ammonium salt to produce an oleophilic clay is commonly employed to obtain a suitable mud.

The oil extracted from a well usually contains water dispersed as a fine emulsion. The de-emulsification of the oil is an important part of the overall processing, and improved and more economical methods of achieving this depend on the application of colloid techniques.

The oil industry is also concerned in many of its processes with the formation of foams which may be either advantageous or undesirable. The control of foaming is therefore another major problem which requires for its solution the application of colloid and surface science.

Finally, the dispersion of oil slicks resulting from spillage has required an understanding of the way in which surfactants can disintegrate the emulsion (*chocolate mousse*) which forms spontaneously by the interaction between crude oil and sea water.

The coal industry also has processes where colloid technology is applied. Thus the elimination of fine coal-dust from wash water is aided by the addition of flocculants such as high-molecular-weight polymers which act by the bridging flocculation mechanism. Although the use of concentrated coal suspensions in oil or water as 'liquid' fuels has been considered and experimented with for at least sixty years, it is only recently that practicable methods have been developed for the preparation and storage of such materials. Here the problem is that of forming a stable, concentrated suspension which has rheological properties that enable it to be pumped and injected into a furnace: the interchangeability of oil-fired and coal-fired operation clearly has many advantages.

In the nuclear industry one of the main applications of colloid technology is in the handling and reprocessing of radioactive waste, where the problems are mainly concerned with the flocculation and separation of particulate radioactive materials. A second application is that of the sol–gel process to the preparation of nuclear fuel in the form of spherical particles of uniform size.

COLLOIDS IN THE FOOD INDUSTRY

Many aspects of colloid science are involved in research on the influence of the methods of processing foods and of additives of various kinds on their rheological and mechanical properties with a view to improving both their handling in manufacture and in the kitchen and their gastronomical quality. As indicated in Chapter 13, gels are present in wide variety in both the preparative stages and final products of the food industry; consequently it is in this area that most interest lies. Traditionally gels were derived from natural biological materials, but synthetic or chemically modified natural products are being used increasingly in manufactured food. Another problem, not confined to the food industry, is that of the wetting of fine powders. Thus it is a matter of some importance that soup, cocoa, and dried-milk powders should be readily dispersible in water. This is achieved by the addition of small amounts of appropriate emulsifying and dispersing agents, usually of biological origin. The control of the particle size of ice crystals in ice-cream and similar frozen products subjected to rapid freezing presents yet another aspect of colloid science of significance in the food industry.

FOAMS

In the case of foams, problems both of the creation of stable foams, important in the fighting of oil fires, and of the destruction of unwanted foams on rivers or at sewage works are of economic and environmental consequence.

Foams for fire-fighting purposes have to meet a number of critical requirements. They must be easily produced in huge volumes and be stable for as long as possible at elevated temperatures, and when they break down it is a great advantage if the aqueous film so produced spreads spontaneously over the oil surface, thus excluding oxygen. Surfactants, many of them fluorochemicals, are now available which in suitable formulations with foaming agents meet many of the desired properties and reduce the enormous risks associated with petrol and oil fires.

On the other hand, great environmental damage and problems in managing sewage treatment plants have been caused by foams produced by the effluent from processes involving surfactants. Antifoaming agents are therefore of major importance in avoiding such problems, which are also mitigated by the use of biodegradable surfactants. Unwanted foams are also produced in the course of many manufacturing operations, and a wide variety of ways of dealing with them have been developed.

The use of foams in mineral flotation processes has been mentioned in Chapter 12. The economic value of this technique can be appreciated from the fact that many millions of tons of ore are treated annually worldwide by flotation processes and that it is now possible to extract economically low-grade ores which a few decades ago were considered useless.

ELIMINATION OF UNWANTED COLLOIDS

We have mentioned above the need to destroy emulsions and foams. Of equal importance in some industries is the need to eliminate particulate dispersions. Thus the final stages of water purification are dependent on the efficient flocculation of particulate matter. This is achieved either by making use of the ability of highly charged ions to cause flocculation or by using a dilute high-molecular-weight polymer to cause bridging floccuation. Since most of the colloidal particles are negatively charged, aluminium ions, Al^{3+}, are the most widely used flocculants; high-molecular-weight polyacrylamides are alternative materials. Yet another aspect of this problem is that of the control of the production of aerosols and fine dusts and of their deposition or filtration. Pollution control, upon which developing emphasis is being placed, is thus an area in which colloid technology has an increasing role to play.

Chapter 15

The Future of Colloid Science

INTRODUCTION

The past few decades have seen major advances in colloid science on both the theoretical and experimental fronts, and there is every reason to believe that further rapid progress will continue.

Early theories of colloidal dispersions, such as the DLVO theory (Chapter 9) and Einstein's theory of viscosity (Chapter 8), were, of necessity, limited in their applicability to very dilute dispersions. They gave general guidance, however, in the search for an understanding of more concentrated dispersions and formed the basis from which more recent progress has evolved.

The need to understand the behaviour of concentrated dispersions, while an intellectual goal in itself, has become increasingly relevant as the industrial exploitation of colloidal systems has developed. This challenge has been taken up and has provided the stimulus for a great deal of recent work, and no doubt it will continue to do so.

This chapter summarises briefly some of these recent developments and indicates the broad areas in which further advances are likely to be made.

VAN DER WAALS FORCES

In Chapter 3 we discussed the calculation of the interaction between two particles by summing the intermolecular potential energies between all the molecules in one particle with all those in the other. This is often called the *Hamaker* or *microscopic theory* and assumes pair-wise additivity of the intermolecular potential energies. This implies that the potential associated with the interactions of a given molecule is independent of the presence of neighbouring molecules. In reality the motions of the electrons in a given atom or molecule, on which the intermolecular interaction depends, are influenced by its neighbours. One should consider each particle as an assembly of molecules whose electronic vibrations are coupled to one another. This coupling is manifested in the dielectric constant and the refractive index of the solid. An alternative theory, the *Lifshitz* or *macroscopic theory*, sets out to relate the forces between the particles to the dielectric properties of the solids and the intervening medium. The solids and the medium are thus regarded as continuous media and their atomic structure is ignored. It turns out that to apply the theory one has to know the dielectric response of the solids and the intervening medium over a very wide range of frequencies. This information is not always readily available, although methods of deriving it from a limited amount of data have been developed. One result of this theory is that the so-called Hamaker constant in fact varies with the separation between the particles, and when used in equation (3.11) it should be regarded as a function of H. It should be called the *Hamaker function*, $A(H)$. The theory enables the numerical values of $A(H)$ to be calculated with some certainty, whereas the Hamaker constant derived from equation (3.12) depends on estimates of α, ν, and q, whose values are not always reliably known.

In broad terms the macroscopic theory does not introduce any major modifications to the basic ideas which have been developed in earlier chapters, although its greater power enables more exact calculations to be made on more complex systems (*e.g.* aqueous systems) than is possible with the microscopic theory. Its main shortcoming is that at very small separations – where the molecular structure of each solid is 'seen' by the other – the theory becomes less reliable and certain somewhat arbitrary approximations have to be made. For the moment we are left with a choice between these two theoretical approaches: which one is chosen depends very largely on the particular problem under consideration and the degree of precision required.

STATISTICAL MECHANICS

Statistical mechanics is relevant to problems in colloid science at two levels. At the molecular level one is concerned in particular with statistical-mechanical theories of electrolyte solutions, the electrical double layer, interfacial regions, and polymer solutions. In these areas developments in basic physical chemistry are of direct interest in colloid science.

The original theory of double-layer interaction dealt with the problem of the repulsion between two isolated, charged colloidal particles and could, strictly, be applied only to very dilute dispersions. More realistic theories must handle concentrated systems where the interaction between many particles has to be considered. This more complex situation has been, and will continue to be, the subject of important theoretical studies.

Of particular significance are recent advances in the study of the molecular structure of interfaces. Thus it is now possible to calculate the form of the local concentration profiles at liquid/vapour and liquid/solid interfaces, leading to detailed theories of adsorption and interfacial tension. These calculations have a special relevance in the development of increasingly realistic theories of steric stabilisation. Thus rapid progress is being made in the prediction of the segment density distributions of polymers adsorbed both at a single interface and on two approaching surfaces; the situations depicted schematically in Figure 3.10 are being given a quantitative description. In addition, the theory is being extended to the case of adsorbed polyelectrolytes.

A second and potentially very significant development arises from the fact that in statistical mechanics the basic theory does not specify the nature of the 'particles' whose behaviour it sets out to describe. In conventional physical chemistry the 'particles' are identified with atoms or molecules, but the fundamental equations apply equally to the behaviour of colloidal particles, as Perrin showed many years ago (see Chapter 6). There is, however, an important difference between molecular and colloidal systems. While in molecular systems the intermolecular forces operate through space and nearest-neighbour interactions dominate, in colloidal systems the forces are of much longer range and are strongly influenced by the nature of the intervening medium. These differences do not, however, obscure the many parallels between theories of the behaviour of colloids and molecular systems, as is clear from the similar approaches we made in discussing flocculation (Chapter 9) and micellisation (Chapter 11).

In addition to analytical theories, the use of computer simulations, in predicting the structure of concentrated dispersions, and of interfacial regions has expanded rapidly and is producing many interesting results for comparison with those obtained experimentally by neutron and light scattering.

The application of statistical theory to kinetic phenomena such as Brownian motion and rheology is also a field in which renewed activity is developing.

In many of these areas progress has only become possible through the availability of powerful high-speed computers, which have enabled work to be done that thirty years ago would have been quite impracticable.

LIGHT SCATTERING

As already indicated in Chapter 7, the introduction of laser technology has already had a major impact on light-scattering methods. These have found particular application in the development of new methods of particle sizing, and several instruments are now available commercially which are designed for the automatic determination of particle size distributions. These methods are being developed steadily, especially in terms of the associated computer software needed for the rapid analysis of experimental data. In particular, while the measurement of the particle size in monodisperse systems is well established, the mathematical analysis for polydisperse systems and for non-spherical particles presents problems which are not yet fully solved.

Both static and dynamic light-scattering techniques will continue to find wide applications in the study of the structure of more concentrated systems.

NEUTRON SCATTERING

Although the general principles involved in neutron scattering are essentially the same as those in the scattering of light, the detailed methods of application are, as mentioned in Chapter 7, rather different. Consequently opportunities for new types of investigation are opened up. Small-angle scattering is also being applied extensively to the study of the structure of concentrated dispersions, and the results are providing experimental tests of the application of statistical mechanics to such systems.

Among other topics that can now be studied are the distribution of polymer segments at an interface, the dynamics of water adsorbed in clays and lamellar liquid crystals, and, using the more recent technique of 'neutron spin echo', the diffusion of colloidal particles and its dependence on particle concentration.

NUCLEAR MAGNETIC RESONANCE

The last two decades have seen a dramatic increase in the applications of nuclear magnetic resonance techniques in colloid and surface science. The reasons for this are two-fold: first, the availability of economical high-field superconducting magnets and, secondly, the development of fast and compact digital computers. These advances have made it possible not only to expand the range of available nuclei that can be observed but also to improve the sensitivity of the method.

In applications to surface and colloid problems it is essential to have both high sensitivity and the ability to mask out signals from the bulk phases. This can be achieved by isotopic substitution (^{2}D for ^{1}H and ^{13}C for ^{12}C) and by selective saturation of unwanted signals using combinations of exciting pulses. By placing both the pulse timing and sequencing under computer control, the range of possible experiments has been greatly increased.

Pulsed NMR techniques provide information on the relaxation times of various molecular motions in the system. Thus, in a very early application, the water adsorbed in silica gel could be divided into a fraction that had been perturbed by adsorption, the rest having properties essentially the same as those of bulk water. Similar information may be obtained on the 'bound' and 'free' water in other gels. The motion and orientation of molecules adsorbed on graphite and clays has also been studied. It is now becoming possible to examine multi-component systems. For example, by measuring the diffusion coefficients of each of the components in microemulsions, it is possible to distinguish between molecules at the interface, solubilised or in the bulk phase. Similar experiments can monitor the percolation of reactants and products in zeolites and supported metal catalysts. In ordered systems such as liquid crystals a wide range of multiple quantum experiments is now available for determining both structure and diffusion.

The technique of NMR imaging, now finding important applications in diagnostic medicine, can also be used in other ways, enabling for example the distribution of oil in rocks or of macroscopic phase transitions in alloys to be detected and studied.

NMR methods can also be used for structural analysis, detecting the local geometries around specific cations in zeolites and so complementing the information obtained by *X*-ray diffraction.

RHEOLOGY

A very direct way of investigating the properties of concentrated dispersions is to use rheological measurements to elucidate the way in which their structure is modified when they are subjected to mechanical stress. Studies of this kind are of special relevance since in many industrial applications it is the rheological behaviour that is of crucial importance. Work in this field has been greatly facilitated in recent years as a result of the development of an increasingly sophisticated range of rheological techniques involving fully automated equipment in which systems can be subjected to a wide range of stress/time regimes. The interpretation of the extensive information obtained in this way is a formidable task which will constitute a continuing theoretical challenge. Research in rheology will certainly represent one of the major growing points in colloid science in the next few decades.

DIRECT FORCE MEASUREMENTS

Theories of interparticle forces play a fundamental part in many theoretical aspects of colloidal behaviour. It is therefore of great importance to have experimental evidence for the validity of these theories. One approach to this is to study the forces between macroscopic objects, to which the same theoretical equations should apply. Since these forces are exceedingly small until the bodies come into very close proximity, work in this area has faced considerable experimental difficulties. Experiments on the force between two plates and between a plate and a lens have been of limited validity because of the difficulty in achieving adequate surface smoothness and in completely eliminating dust. The closest distances of approach that can be studied in such configurations are several tens of nanometres. Major successes have resulted from the discovery that freshly cleaved mica surfaces are smooth on the atomic scale and from the fact that the force between crossed cylindrical specimens (Figure 15.1) follows the same law as that between a sphere and a plate. Sophisticated techniques for bringing such surfaces together to within precisely known distances and measuring the resulting force have been developed in recent years and are still being improved. It is now possible to make accurate measurements down to separations of a few tenths of a nanometre (a few ångströms). Initial work on forces in air has been extended to mica surfaces immersed in pure liquids, in electrolyte solutions, and in polymer solutions. Evidence on the validity of the various theories of interparticle interactions is now accumulating. Work of this kind is destined to make a major contribution to the ultimate understanding of colloidal phenomena.

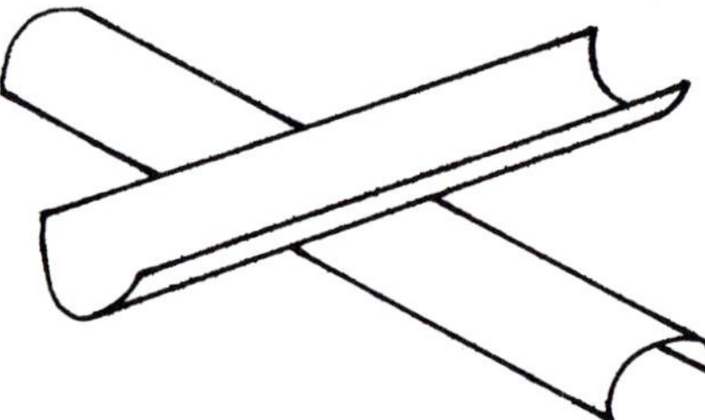

Figure 15.1 *Arrangement of crossed cylinders as used for the direct determination of the force between macroscopic bodies.*

BIOLOGICAL SYSTEMS

The importance of the role played by colloids in biological systems is becoming increasingly understood. Colloidal properties and structures are significant in a wide range of phenomena, including, as a few representative examples, platelet and cell adhesion, membrane transport, phagocytosis, blood rheology, immunology, bone regeneration, and photosynthesis. In these and other areas rapid progress is being made towards an understanding of the fundamental processes concerned, and this is likely to continue.

CONCLUSIONS

In this chapter we have mentioned only a few of the more important future developments which can be foreseen in colloid science. Many of these will depend on the availability of modern instrumentation and of powerful computer facilities. In addition to the techniques dealt with in this chapter, mention should also be made of the contributions from greatly improved electron microscopic techniques, ultracentrifuges, and *X*-ray equipment. Other techniques that will become of increasing significance include dielectric measurements, electrical birefringence, and time-resolved fluorescence.

The armoury of the colloid scientist is expanding year by year and leading to an ever-increasing understanding of the phenomena with which he deals, but the fundamental principles upon which this depends are unlikely to change.

Appendix I

Preparation of Some Simple Colloidal Systems

It is particularly important in all work on colloids to observe the most stringent precautions to ensure the absolute cleanliness of all apparatus.

Gold Sol

Add 1 cm^3 of a 1% solution of gold chloride ($HAuCl_4 \cdot 3H_2O$) to 100 cm^3 of distilled water, bring to the boil, and add 2.5 cm^3 of 1% sodium citrate solution. Keep the solution just boiling. After a few minutes observe the appearance of a blue coloration, followed shortly by the formation of a ruby-red gold sol.

Sulphur Sol

Mix rapidly in the cold equal amounts of 0.004 mol dm^{-3} sodium thiosulphate solution and 0.004 mol dm^{-3} hydrochloric acid. The mixture becomes cloudy after a few minutes and then develops to an opaque white dispersion of colloidal sulphur.

Silver Bromide Sol

Mix equal amounts of 0.020 mol dm^{-3} sodium bromide solution and 0.018 mol dm^{-3} silver nitrate solution. A colloidal dispersion of silver bromide is formed immediately. A silver iodide sol may be prepared in a similar manner.

Ferric Hydroxide Sol

Add 2 cm^3 of a 30% solution of ferric chloride slowly, with stirring, to 500 cm^3 of boiling distilled water. A clear reddish-brown dispersion of ferric hydroxide is formed.

Emulsions and Microemulsions

A *simple emulsion* can be prepared by shaking a solution of household liquid detergent (washing-up liquid) diluted 1:10 with an equal volume of white spirit (turpentine substitute) or a few drops of cooking oil. The resulting emulsion is stable for several hours.

Microemulsions can be produced in several ways. One simple recipe is to use 7 cm^3 of household liquid detergent (neat or diluted 1:2), 14 cm^3 of white spirit, and 4 cm^3 of n-pentanol (amyl alcohol) or n-butanol. The components are mixed gently and allowed to separate into two phases. After several hours the upper oil phase becomes clear but (see Appendix II) exhibits a strong Tyndall cone. This indicates the presence of microemulsion droplets of colloidal size.

Polymer Latex

The preparation of polymer latices is a more lengthy process and requires more specialised apparatus and laboratory facilities. A student experiment for the preparation of polystyrene latex by emulsion polymerisation (see Chapter 4) is described by M. W. J. Esker and J. H. A. Pieper in 'Physical Chemistry: Enriching Topics from Colloid and Surface Chemistry', ed. H. van Olphen and K. J. Mysels, Theorex, California, 1975.

Appendix II

Some Simple Experiments with Colloids

(1) Observe that both the gold sol and the ferric hydroxide sol appear quite clear when viewed in ordinary light. Now pass a beam of light through each of them and observe a Tyndall cone. Suitable light sources are a beam from a slide projector* or, in a darkened room, a beam from a hand torch. Similar observations can be made with diluted samples of the sulphur and silver bromide sols and with microemulsions.

(2) Observe a similar Tyndall cone by shining a beam through a developing sulphur sol. If a low-power He/Ne laser is available, observe the scintillations during the formation of the sol (Chapter 7).

(3) Add a little 1 $mol\,dm^{-3}$ sodium chloride to each of the sols and observe whether flocculation occurs. In the case of gold sol flocculation is marked by a change in colour from ruby-red to blue. If the salt solution is added from a burette, the critical flocculation concentration (Chapter 9) can be determined. Repeat using 0.1 $mol\,dm^{-3}$ sodium sulphate solution.

(4) Add 1 cm^3 of a 0.1% solution of gelatine to about 10 cm^3 sol and repeat the above flocculation experiments. Observe the 'protective action' of gelatine.

(5) Mix 1 cm^3 of gold sol with 9 cm^3 of ferric hydroxide sol, shake, and observe for half an hour. The formation of a floc illustrates heteroflocculation (Chapter 9).

* A narrow beam is conveniently obtained by inserting a blank opaque slide containing a central pin-hole.

Appendix III

Definitions and Measurement of Adsorption

THE RELATIVE ADSORPTION

In Chapter 5 the surface excess of component i was defined in equation (2.1) as

$$n_i^\sigma = n_i - c_i^\alpha V^\alpha - c_i^\beta V^\beta. \tag{AIII.1}$$

As pointed out there, one has to devise some means of defining V^α and V^β unambiguously. We may begin by choosing to define the boundary between α and β by a dividing surface placed arbitrarily at a level z^δ [Figure 5.3(b)]. The adsorption on unit area of surface is given by the sum of the shaded areas in this figure, and this is clearly dependent on the choice of z^δ. However, it turns out that, when the adsorptions of several components of the solution are all defined with respect to the *same* dividing surface, there is a simple relationship between them.

This is easily seen by writing down the equation for the surface excess of, say, component 1:

$$n_1^\sigma = n_1 - c_1^\alpha V^\alpha - c_1^\beta V^\beta. \tag{AIII.2}$$

In addition we have also

$$V = V^\alpha + V^\beta, \tag{AIII.3}$$

where V is the total volume. There are now three equations involving V^α and V^β, so that by simple algebra they can be eliminated. If we denote $(c_i^\beta - c_i^\alpha)$ by Δc_i, we can write the resulting equations in the form

$$\frac{n_1^\sigma - n_1 + c_1^\alpha V}{\Delta c_1} = \frac{n_2^\sigma - n_2 + c_2^\alpha V}{\Delta c_2} = \frac{n_3^\sigma - n_3 + c_3^\alpha V}{\Delta c_3} \ldots \tag{AIII.4}$$

Taking any of the right-hand members, equating to the first, and rearranging gives

$$n_i^\sigma - n_1^\sigma(\Delta c_i/\Delta c_1) = [n_i - c_i^\alpha V] - [n_1 - c_1^\alpha V](\Delta c_i/\Delta c_1). \tag{AIII.5}$$

Now all the quantities on the right-hand side of equation (AIII.5) are experimentally measurable, so that the expression on the left must be independent of the choice of the dividing surface: it is called the *relative surface excess of i with respect to component 1* and is denoted by $n_i^{\sigma(1)}$:

$$n_1^{\sigma(1)} = n_i^\sigma - n_1^\sigma(\Delta c_i/\Delta c_1). \tag{AIII.6}$$

One is often interested in the surface excess divided by the surface area (A_s):

$$\Gamma_i^{(1)} = n_i^{\sigma(1)}/A_s = \Gamma_i - \Gamma_1(\Delta c_i/\Delta c_1), \tag{AIII.7}$$

called the *relative adsorption of component i* (or the *relative areal surface excess of i*) with respect to component 1. In equation (AIII.7) Γ_i and Γ_1 are defined with respect to an arbitrary dividing surface, but it is clear from equations (AIII.6) and (AIII.7) that, had we chosen the surface for which n_1^σ or $\Gamma_1 = 0$, then $n_i^{\sigma(1)}$ and Γ_i would have been the values corresponding to this choice. It is for this reason that a common definition of the relative adsorption relates to the choice of the dividing surface for which $\Gamma_1 = 0$.

MEASUREMENT OF ADSORPTION

Liquid/Vapour Interface

The direct measurement of adsorption in this case is not always easy. Thus for a system containing a liquid/vapour interface the two terms in equation (AIII.5) are nearly equal and a very highly precise analytical technique is needed to obtain reasonably accurate results. One way in which this problem can be reduced is seen by rewriting equation (AIII.5), for a binary solution, and neglecting the concentration in the vapour phase:

$$n_2^{\sigma(1)} = V^l c_i^l[\{n_2/(V^l c_2^l)\} - \{n_1/(V^l c_1^l)\}]. \tag{AIII.8}$$

If V^l is very small, then, since for a given interfacial area $n_2^{\sigma(1)}$ is independent of the amount of liquid taken, the term in square brackets must be proportionally larger and hence measureable with higher accuracy. This equation was the basis of McBain's attempts to measure $\Gamma_2^{(1)}$ experimentally, the small values of V^l being achieved by slicing a thin layer from the surface of a solution of known concentration, c_1^l and c_2^l, using a fast-moving microtome. Analysis of the liquid sample so collected enabled n_2 and n_1 to be obtained. Other similar methods have been tried, but direct analytical measurements rarely give results accurate to better than a few percent. In this instance, therefore, one has to rely on the indirect determination of $\Gamma_2^{(1)}$ through measurements of the surface tension of solutions making use of the Gibbs adsorption equation.

Similar considerations apply to the liquid/liquid interface.

Liquid/Solid Interface

Here the problem is very much simpler, mainly because the surface areas of particulate or porous solids are much higher than the available areas of liquid/vapour interfaces. If no component of the solution penetrates into the solid (α), then $c_i^\alpha = 0$ and equation (AIII.5) can be written, for a binary solution, in the form

$$n_2^{\sigma(1)} = n_2 - n_1(c_2^l/c_1^l) = n_2 - n_1(x_2^l/x_1^l) = (n_2 x_1^l - n_1 x_2^l)/x_1^l, \tag{AIII.9}$$

where x_1^l and x_2^l are the mole fractions in the liquid. In an experimental determination of adsorption from solution a sample of solution of initial concentration x_1^0 and x_2^0, containing a total amount n^0, is contacted with a mass m of solid, and when equilibrium is achieved the final concentration x_2^l is measured. The total amounts of 1 and 2 present are, respectively, $n^0 x_1^0$ and $n^0 x_2^0$. Substituting in equation (AIII.9), remembering that $x_1 + x_2 = 1$, gives

$$n_2^{\sigma(1)} = n^0(x_2^0 - x_2^l)/x_1^l = n^0 \Delta x_2^l/x_1^l, \tag{AIII.10}$$

where Δx_2^l is the change in the mole fraction of 2 in the solution resulting from adsorption at the solid surface. The surface excess for unit mass of solid is called the *specific relative surface excess*,

$$n_2^{\sigma(1)}/m = n^0 \Delta x_2^l/m x_1^l, \tag{AIII.11}$$

while, if the specific surface area of the solid (a_s) is known, one can define the *areal relative adsorption* as

$$\Gamma_2^{(1)} = (n^0 \Delta x_2^l / m a_s)/x_1^l. \tag{AIII.12}$$

The meaning of the numerator of this expression is seen by considering the difference between the amount of component 2 actually present ($n^0 x_2^0$) and the amount that would have been present if the equilibrium concentration had been uniform in the liquid up to the solid surface ($n^0 x_2^l$). This is the surface excess corresponding to the choice of the solid surface as the dividing surface:

$$n_2^{\sigma(n)} = n^0 \Delta x_2^l, \tag{AIII.13}$$

and is called the *reduced surface excess*. The specific reduced surface excess and the areal reduced surface excess are defined, respectively, by division by m and ma_s.

It follows that

$$\Gamma_2^{(1)} = \Gamma_2^{(n)}/x_1^l. \tag{AIII.14}$$

In dilute solution ($x_1^l \rightarrow 1$) the relative and reduced surface excesses become equal.

Experimental data for adsorption from solution are usually expressed as specific reduced surface excess isotherms, sometimes called *composite isotherms*, in which $n^0 \Delta x_2^l / m$ is plotted as a function of x_2^l.

Appendix IV

The Gibbs Adsorption Equation

It is well known that to define unambiguously the state of a bulk system it is necessary to specify the values of a certain number of experimental variables. For a one-phase system (α) we may, for example, specify the values of T^α, V^α, and n_i^α ($i = 1, \ldots c$). This is the most appropriate choice when employing the Helmholtz free energy (F) of the system. It follows that the total differential of F may be written

$$\mathrm{d}F = -S^\alpha \mathrm{d}T^\alpha - p^\alpha \mathrm{d}V^\alpha + \sum_l \mu_i^\alpha \mathrm{d}n_i^\alpha. \qquad \text{(AIV.1)}$$

A similar equation applies to phase β. If the phases exist in equilibrium, with a plane interface between them,

$$T^\alpha = T^\beta,\ p^\alpha = p^\beta, \text{ and } \mu_i^\alpha = \mu_i^\beta \text{ (all } i). \qquad \text{(AIV.2)}$$

The presence of an interface between the phases influences the free energy of the whole system, which is therefore given by

$$F = F^\alpha + F^\beta + F^\sigma, \qquad \text{(AIV.3)}$$

where F^σ is the surface free energy. In conventional bulk thermodynamics, where the interfacial areas are small, F^σ is ignored. For colloidal systems, however, F^σ plays an important role.

We have again to divide the system into the phases α and β, *i.e.* define the volumes V^α and V^β. Adopting the device of a dividing surface, it turns out that the total differential of the surface free energy is

$$\mathrm{d}F^\sigma = -S^\sigma \mathrm{d}T + \sigma \mathrm{d}A_\mathrm{s} + \sum_i \mu_i \mathrm{d}n_i^\sigma, \qquad \text{(AIV.4)}$$

where the μ_i's have the same values as in the two adjoining phases. The term $\sigma \mathrm{d}A_\mathrm{s}$ replaces the $p\mathrm{d}V$ term (with the opposite sign since σ is a tension not a pressure) in the equations for the bulk phases.

In bulk thermodynamics one derives the *Gibbs–Duhem equation* by integrating equation (AIV.1), keeping the intensive properties constant ($dT = 0$, p and μ_i constant) to obtain

$$F^\alpha = pV^\alpha + \sum_i \mu_i^\alpha n_i^\alpha. \qquad \text{(AIV.5)}$$

This is differentiated generally, and the expression for dF thus obtained is equated to that in equation (AIV.1) to yield the Gibbs–Duhem equation:

$$V^\alpha dp - S^\alpha dT + \sum_i n_i^\alpha d\mu_i^\alpha = 0. \qquad \text{(AIV.6)}$$

An exactly analogous procedure is used to derive the *Gibbs adsorption equation*. On integration of equation (AIV.4) with T, σ, and μ_i constant we find

$$F^\sigma = \sigma A_s + \sum_i \mu_i n_i^\sigma. \qquad \text{(AIV.7)}$$

Differentiation and subtraction from equation (AIV.4) gives

$$-A_s d\sigma - S^\sigma dT + \sum_i n_i^\sigma d\mu_i = 0. \qquad \text{(AIV.8)}$$

This is the Gibbs adsorption equation. It is usually applied at constant temperature; when expressed in terms of the adsorption it becomes the Gibbs adsorption isotherm:

$$-d\sigma = \sum_1^c \Gamma_i d\mu_i. \qquad \text{(AIV.9)}$$

Up to this point we have not specified the position of the dividing surface. If we now choose this so that $\Gamma_1 = 0$, then

$$-d\sigma = \sum_2^c \Gamma_i^{(1)} d\mu_i, \qquad \text{(AIV.10)}$$

where the summation no longer involves $d\mu_1$. An alternative derivation follows from the argument that from the phase rule it is not possible at constant T and p to vary all the μ_i's simultaneously. This is expressed in the Gibbs–Duhem equation. One can therefore use the Gibbs–Duhem equations for the two bulk phases to eliminate $d\mu_1$ from equation (AIV.9) and obtain equation (AIV.10).

Appendix V

Influence of Adsorption on Interparticle Forces

The discussion given in Chapter 5, equations (5.9)—(5.12), may be taken one stage further in the following way.

Equation (5.12) may be integrated to give

$$\mathscr{F}(H) - \mathscr{F}_1^*(H) = 2\int_{\mu_2=-\infty}^{\mu_2} (\partial\Gamma_2^{(1)}/\partial H)\mathrm{d}\mu_2. \qquad \text{(AV.1)}$$

Here $\mathscr{F}(H)$ is the force between the plates in a solution of mole fraction x_2^l, while $\mathscr{F}_1^*(H)$ is that when the plates are in pure component 1. Now in component 1 the force arises solely from van der Waals forces, so that the right-hand side represents the contribution of adsorption to the interparticle force.

Equation (AV.1) can be written in terms of the mole fraction of 2 in the solution:

$$\mathscr{F}(H) - \mathscr{F}_1^*(H) = 2\int_{x_2=0}^{x_2} (\partial\Gamma_2^{(1)}/\partial H)\mathrm{d}\ln x_2^l\gamma_2^l, \qquad \text{(AV.2)}$$

where γ_2^l is the activity coefficient of component 2 at the mole fraction x_2^l.

The qualitative generalisations stated in Chapter 5 follow immediately. To derive a quantitative result it is necessary to know, or have a theory to predict, the way in which the adsorption isotherm depends on the plate separation.

Appendix VI

Steric Stabilisation

Many modern theories of polymer solutions have in common a parameter (usually denoted by χ) which measures the difference between the interaction energy (w_{12}) between a molecule of solvent and a polymer segment and the arithmetic mean $[(w_{11} + w_{22})/2]$ of the solvent–solvent (w_{11}) and segment–segment (w_{22}) interaction energies (Figure AVI.1):

$$\chi = [w_{12} - (w_{11} + w_{22})/2]z/kT, \qquad \text{(AVI.1)}$$

where z is the number of nearest neighbours of a segment or solvent molecule in solution and where the volumes of solvent molecule and polymer segment are taken to be the same. Since, according to simple approximate theories of London forces, $|w_{12}|$ is the geometric mean of w_{11} and w_{22},

$$w_{12} = -(w_{11}w_{22})^{1/2}, \qquad \text{(AVI.2)}$$

χ is expected to be positive.

The configuration of polymer molecules in solution is determined by the value of χ. When χ is large, segment–solvent interactions are relatively weak and the configuration is dominated by segment–segment interactions leading to a compact coil configuration ($\alpha < 1$ in Figure 3.10); when χ is small, the polymer is in the form of an extended chain ($\alpha > 1$). The critical value of χ is $\frac{1}{2}$ and corresponds to a random coil configuration. This is identified with the so-called θ-configuration ($\alpha = 1$).

Similar considerations are important in theories of polymer adsorption, although here the relative magnitudes of surface–solvent and surface–segment interaction energies have to be taken into account.

When applied to the problem of steric stabilisation, many theories predict that the resulting interaction energy between the

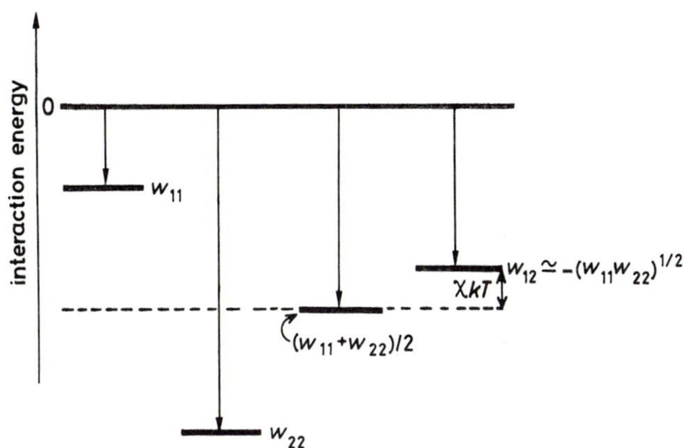

Figure AVI.1 *Relationship between solvent–solvent, segment–segment, and solvent–segement interaction energies, and the definition of the interaction parameter, X, for polymer solutions.*

particles should be proportional to $(\frac{1}{2} - \chi)$. For $\chi < \frac{1}{2}$ there is a steric repulsion, while for $\chi > \frac{1}{2}$ steric effects contribute an additional attractive term. On this basis, if one is justified in comparing the behaviour of an adsorbed polymer with that in solution, one would expect to find a relation between the critical flocculation conditions, brought about by changing the medium or the temperature, and the θ-point of the bulk polymer solution. This is often observed, although other factors can upset this simple prediction, especially when the surface–segment attraction energy is large. In the case of block copolymers the correlation may be expected with the θ-condition of the 'soluble' chain of the copolymer.

Appendix VII

Further Reading

Relatively few general books on colloid science are currently in print, so that the reader wishing to extend his knowledge may find it necessary to make use of library facilities.

The following is not an exhaustive bibliography but includes both relatively elementary books and a selection of more advanced texts and papers.

General

Among the classical works on colloids, several of which give a good historical perspective, are:

E. Hatschek, 'The Foundations of Colloid Chemistry', Ernest Benn, London, 1925. This book contains reprints and translations of papers by: Ascherson, 1840; van Bemmelen, 1888; Faraday, 1857; Graham, 1864; Carey-Lea, 1889; Muttmann, 1887; Selmi, 1845.

H. Freundlich, 'Colloid and Capillary Chemistry', 1st Engl. Edn., Methuen, London, 1926.

E. S. Hedges, 'Colloids', Edward Arnold, London, 1931.

A. von Buzagh, 'Colloid Systems', Technical Press, London, 1937.

A. E. Alexander and P. Johnson, 'Colloid Science', Cambridge University Press, 1949, 2 Vols.

'Colloid Science', ed. H. R. Kruyt, Elsevier, Amsterdam, London, New York, 1949, 2 Vols.

More recent shorter books include:

A. G. Ward, 'Colloids', Blackie and Son, London, Glasgow, 1945.

K. J. Mysels, 'Introduction to Colloid Chemistry', Interscience, New York, 1959.

M. J. Vold and R. D. Vold, 'Colloid Chemistry', Chapman and Hall, London, 1965.

G. D. Parfitt, 'Principles of the Colloid State', Monographs for Teachers, No. 14, Royal Institute of Chemistry, London, 1967.

W. J. Popiel, 'Introduction to Colloid Science', Exposition Press, Hicksville, New York, 1978.

D. J. Shaw, 'Introduction to Colloid and Surface Chemistry', 3rd Edn., Butterworths, London, 1980.

A broad survey of a number of interesting aspects of surface chemistry and its relationship to colloid science is presented in:

The Nuffield Advanced Science Study Option on Surface Chemistry, Students' Book and Teachers' Guide, Longman, 1984.

Papers presented at a symposium on the teaching of colloid and surface chemistry held in 1961 are published in:

'The Teaching of Colloid and Surface Chemistry', in *J. Chem. Educ.*, 1962, **39**, 166.

Material presented at a Royal Society of Chemistry review symposium on colloid science held in 1981 is published in:

'Colloidal Dispersions', ed. J. W. Goodwin, Special Publication No. 43, The Royal Society of Chemistry, London, 1982. (This title is referred to as 'Colloidal Dispersions' in the following references.)

A stimulating book which emphasises the link between physical chemistry and colloid science is:

'Physical Chemistry: Enriching Topics from Colloid and Surface Science', ed. H. van Olphen and K. J. Mysels, IUPAC Commission I.6, Theorex, La Jolla, California, 1975. (This book is abbreviated to 'Enriching Topics' in subsequent references.)

The main modern advanced treatises are:

P. C. Hiementz, 'Principles of Colloid and Surface Chemistry', 2nd Edn., Dekker New York, 1985.

R. J. Hunter, 'Foundations of Colloid Science', Oxford University Press, 1987, Vol. 1; Vol. 2 in press.

Many excellent review articles are given in:

Surface and Colloid Science Series, ed. E. Matijevic, Wiley–Interscience, New York, London, Sydney, Toronto, 1969— .

Advances in the period 1970—1981 are reviewed in:

'Colloid Science', ed. D. H. Everett (Specialist Periodical Reports), The Chemical Society, 1973—1979, Vols. 1—3; The Royal Society of Chemistry, 1983, Vol. 4.

Chapter 1

General brief introductions to colloid science are to be found in the following:

E. Matijevic, 'Colloids: the World of Neglected Dimensions', *Chem. Technol.*, 1973, **3**, 656.

J. A. Kitchener, 'Colloids: New Life from Old Roots', *Chem. Br.*, 1977, **13**, 105.

R. H. Ottewill, 'Colloid Chemistry – Today and Tomorrow', *Prog. Colloid Polym. Sci.*, 1976, **59**, 14.

D. H. Everett, 'Colloid Science – or is it?', *Proc. R. Inst.*, 1978, **52**, 209.

J. Th. G. Overbeek, 'Colloids, a Fascinating Subject', in 'Colloidal Dispersions', Ch. 1.

D. H. Everett, 'Colloids in the World Around Us', *Chem. Br.*, 1981, **17**, 377.

Chapter 2

A general article on the stability of colloidal systems is given by H. van Olphen in 'Enriching Topics', Ch. 2.

Chapter 3

Intermolecular forces and those between macroscopic bodies are treated in more detail and the origin of the stability of colloids is outlined in:

D. Tabor, 'Gases, Liquids and Solids', 2nd Edn., Cambridge University Press, 1979, pp. 16—29.

A more extensive discussion of attractive forces is given in:

D. Tabor, in 'Colloidal Dispersions', Ch. 2.

An advanced treatise is:

J. N. Israelachvili, 'Intermolecular and Surface Forces', Academic Press, London, 1985.

Equation (3.18) illustrates the application of the Boltzmann Law to the distribution of molecules in the atmosphere, assuming the temperature is constant throughout. What happens when this is not so is discussed in:

L. K. Nash, 'Energy Distributions that Deny Boltzmann', *J. Chem. Educ.*, 1984, **61**, 22.

The authoritative book on the electrical double layer is:

M. J. Sparnaay, 'The Electrical Double Layer', Pergamon, Oxford, 1972.

The role of double layers in colloid science is outlined in:

D. Stigter, 'Electrostatic Interactions in Aqueous Environments', in 'Enriching Topics', Ch. 12.

J. Lyklema, 'Electrochemistry of Reversible Electrodes and Colloidal Particles', in 'Enriching Topics', Ch. 19.

J. Lyklema, 'Fundamentals of Electrical Double Layers in Colloidal Systems', in 'Colloidal Dispersions', Ch. 3.

J. Lyklema, 'Electrical Double Layers on Oxides', *Chem. Ind. (London)*, 1987, 741.

More details on steric stabilisation are to be found in:

D. H. Napper, 'Polymeric Stabilisation', in 'Colloidal Dispersions', Ch. 5.

See also references for Chapter 9.

Chapter 4

J. Th. G. Overbeek, 'Monodisperse Systems, Fascinating and Useful', *Adv. Colloid Interface Sci.*, 1982, **15**, 251.

'Polymer Colloids I', ed. R. M. Fitch, Plenum Press, New York, London, 1971.

'Polymer Colloids II', ed. R. M. Fitch, Plenum Press, New York, London, 1980.

R. Buscall, T. Corner, and J. F. Stageman, 'Polymer Colloids', Elsevier, Amsterdam, Oxford, New York, 1985.

Chapter 5

The classical work on surfaces is:

N. K. Adam, 'The Physics and Chemistry of Surfaces', 3rd Edn., Oxford University Press, 1941.

More recent books are:

J. T. Davies and E. K. Rideal, 'Interfacial Phenomena', Academic Press, New York, London, 1961.

R. Aveyard and D. A. Haydon, 'An Introduction to the Principles of Surface Chemistry', Cambridge University Press, 1973.

A. W. Adamson, 'Physical Chemistry of Surfaces', 4th Edn., Wiley–Interscience, New York, London, Sydney, Toronto, 1982.

'Adsorption from Solution', ed. R. H. Ottewill, C. H. Rochester, and A. L. Smith, Academic Press, London, 1983.

'Adsorption from Solution at the Solid/Liquid Interface', ed. G. D. Parfitt and C. H. Rochester, Academic Press, London, 1983.

The thermodynamics of surfaces is dealt with in detail in:

R. Defay, I. Prigogine, A. Bellemans, and D. H. Everett, 'Surface Tension and Adsorption', Longmans, London, 1966.

Chapter 6

A fuller account of the theory of Brownian motion will be found in:

D. Tabor, 'Gases, Liquids and Solids', 2nd Edn., Cambridge University Press, 1979, pp. 112—116.

Einstein's original work is available in:

'Investigations on the Theory of Brownian Motion', Dover, New York, 1956.

Historical accounts of earlier work are given in:

D. Layton, 'The Original Observations of Brownian Motion', *J. Chem. Educ.*, 1965, **42**, 367.

M. Kerker, 'Brownian Motion and Molecular Reality Prior to 1900', *J. Chem. Educ.*, 1974, **51**, 764.

An undergraduate laboratory experiment on Brownian motion is described in:

G. P. Matthews, 'Brownian Motion', *J. Chem. Educ.*, 1982, **59**, 246.

Some recent developments in the theory of Brownian motion are surveyed in:

B. H. Lavenda, 'Brownian Motion', *Sci. Am.*, 1985, **252**, (2), 56.

E. Dickenson, 'Brownian Dynamics and Aggregation: from Hard Spheres to Proteins', *Chem. Ind. (London)*, 1986, 158.

Electrophoresis is dealt with in:
D. Stigter, in 'Enriching Topics', Ch. 20.
R. J. Hunter, 'The Zeta-Potential in Colloid Science', Academic Press, London, 1981.

Chapter 7

A popular account of 'Blue Skies and the Tyndall Effect' is given in:
M. Kerker, *J. Chem. Educ.*, 1971, **48**, 389.

The authoritative books are:
M. Kerker, 'The Scattering of Light', Academic Press, New York, London, 1969.
B. J. Berne and R. Pecora, 'Dynamic Light Scattering', Wiley–Interscience, New York, 1975.

The principles of photon correlation spectroscopy are explained in:
K. J. Randle, 'Statistical Optics and Its Application to Colloid Science', *Chem. Ind. (London)*, 1980, 74.

More recent work on light scattering and neutron scattering is reviewed in:
P. N. Pusey, in 'Colloidal Dispersions', Ch. 6.
R. H. Ottewill, in 'Colloidal Dispersions', Ch. 7.

Chapter 8

The preparation of bouncing putty is described in:
D. A. Armitage *et al.*, *J. Chem. Educ.*, 1973, **50**, 434.

Other useful references are:
J. D. Ferry, 'Rheology in the World of Neglected Dimensions', *J. Chem. Educ.*, 1961, **38**, 110.
J. W. Goodwin, 'Some Uses of Rheology in Colloid Science', in 'Colloidal Dispersions', Ch. 8.
D. C.-H. Cheng, 'Viscosity–Concentration Equations and Flow Curves for Suspensions', *Chem. Ind. (London)*, 1980, 403.
Th. F. Tadros, 'Rheology of Concentrated Suspensions', *Chem. Ind. (London)*, 1985, 210.

Chapter 9

J. Th. G. Overbeek, 'Recent Developments in the Understanding of Colloid Stability', *J. Colloid Interface Sci.*, 1977, **58**, 408.

J. Th. G. Overbeek, 'The Rule of Schultze and Hardy', *Pure Appl. Chem.*, 1980, **52**, 1151.

B. Vincent, 'The Stability of Particulate Suspensions', *Chem. Ind. (London)*, 1980, 218.

D. H. Napper, 'Polymeric Stabilisation', in 'Colloidal Dispersions', Ch. 5.

Th. F. Tadros, 'Polymer Adsorption and Dispersion Stability', in 'The Effect of Polymers on Dispersion Stability', Academic Press, London, 1982.

'Solid–Liquid Dispersions', ed. Th. F. Tadros, Academic Press, London, 1987.

Chapter 10

S. J. Gregg, 'The Surface Chemistry of Solids', 2nd Edn., Chapman and Hall, London, 1961, Ch. 3, Ch. 7.

See also references for Ch. 12.

Chapter 11

The literature on surfactants and micelle formation is extensive. Among the more recent books are:

Micellisation, Solubilisation and Microemulsions', ed. K. L. Mittal, Plenum Press, New York, 1977, Vols. 1 and 2.

'Solution Chemistry of Surfactants', ed. K. L. Mittal, Plenum Press, New York, 1979, Vols. 1 and 2.

'Solution Behaviour of Surfactants', ed. K. L. Mittal, Plenum Press, New York, 1982, Vols. 1 and 2.

'Surfactants in Solution', ed. K. L. Mittal and B. Lindman, Plenum Press, New York, 1984, Vols. 1—3.

'Surfactants in Solution', ed. K. L. Mittal and P. Bothorel, Plenum Press, New York and London, 1986, Vols. 4—6.

'Cationic Surfactants', ed. E. Jungeman, Dekker, New York, 1967.

'Anionic Surfactants', ed. E. H. Lucassen-Reynders, Dekker, New York, 1981.

'Non-Ionic Surfactants', ed. M. J. Schick and J. Cross, Dekker, New York, 1987.

P. H. Elworthy, A. T. Florence, and C. B. Macfarlane, 'Solubilisation by Surface Active Agents', Chapman and Hall, London, 1968.

'Surfactants', ed. Th. F. Tadros, Academic Press, London, 1984.

C. Tanford, 'The Hydrophobic Effect, Formation of Micelles and Biological Membranes', 2nd Edn., Wiley, New York, 1980.

S. E. Friberg and B. Bendikson, 'A Simple Experiment Illustrating the Structure of Associated Colloids', *J. Chem. Educ.*, 1979, **56**, 553.

R. Aveyard, 'Ultralow Surface Tensions and Microemulsions', *Chem. Ind. (London)*, 1987, 474.

D. H. Everett, 'Some Thermodynamic Aspects of Aggregation Phenomena, *Colloids Surf.*, 1986, **21**, 41.

Chapter 12

In addition to the books by Boys and Isenberg mentioned in the text the following references are useful:

J. A. Kitchener, 'Surface Forces in Thin Films', *Endeavour,* 1963, **22**, No. 87, 118.

K. J. Mysels, K. Shinoda, and S. Frankel, 'Soap Films', Pergamon, London, 1969.

J. Th. G. Overbeek, 'Soap Films as a Central Theme in Detergent Research', in *Proc. IVth Int. Congr. Surf. Active Substances,* 1964, Gordon and Breach, London, New York, Paris, 1967, Vol. II, p. 19.

W. H. Slabaugh, 'Soap Bubbles and Flotation', in 'Enriching Topics', Ch. 6.

R. Aveyard and B. Vincent, 'Liquid–Liquid Interfaces: in Isolation and in Interaction', *Prog. Surf. Sci.,* 1977, **8**, 59.

A. W. Adamson, 'Physical Chemistry of Surfaces', 4th Edn., Wiley–Interscience, New York, London, Sydney, Toronto, 1982, Ch. XII. This chapter deals with emulsions and foams and, *inter alia*, with soap films.

P. Becher, 'Encyclopaedia of Emulsion Technology', Dekker, New York, 1983, Vol. 1; 1985, Vol. 2; 1987, Vol. 3.

'Non-Aqueous Foams' (symposium papers), *Chem. Ind. (London),* 1981, 47—60.

J. H. Aubert, A. M. Kraynik, and P. B. Rand, 'Aqueous Foams', *Sci. Am.*, 1986, **254**, (5), 58.

Chapter 13

The subject of gels is treated very inadequately in most more recent texts, but a very full account, though now somewhat outdated, is given in:

P. H. Hermans, in 'Colloid Science', ed. H. R. Kruyt, Elsevier, Amsterdam, London, New York, 1949, Vol. II, Ch. XII, pp. 483—651.

T. Tanaka, 'Gels', *Sci. Am.*, 1981, **244**, (1), 110.

V. J. Morris, 'Food Gels – Role Played by Polysaccharides', *Chem. Ind. (London)*, 1985, 159. This provides some more recent information on the structure of gels.

Gelation phenomena in clay systems are discussed in:

H. van Olphen, 'An Introduction to Clay Colloid Chemistry', 2nd Edn., Wiley–Interscience, New York, London, Sydney, Toronto, 1977.

Chapter 14

General References on Industrial Aspects of Colloids

J. Th. G. Overbeek, 'Technical Applications of Colloid Science', in 'Emulsions, Latices, Dispersions', ed. P. Becher and M. N. Yudenfreund, Dekker, New York, 1978, p. 1.

J. L. Moilliet, B. Collie, and W. Black, 'Surface Activity', 2nd Edn., Spon, London, 1961.

'Dispersion of Powders in Liquids', 3rd Edn., ed. G. D. Parfitt, Applied Science Publishers, London, 1981.

'Industrial Applications of Surfactants', ed. D. R. Karsa, Special Publication No. 59, The Royal Society of Chemistry, London, 1987.

Applications to Specific Industries

D. W. J. Osmond, 'The Impact of Colloid Science on the Paint Industry', *Chem. Ind. (London)*, 1974, 891.

'Paints and Surface Coatings', ed. R. Lambourne, Ellis Horwood, Chichester, 1987.

Th. F. Tadros, 'Control and Assessment of the Physical Stability of Pesticidal Suspension Concentrates', *Chem. Ind. (London)*, 1980, 211.

J. A. Kitchener, 'Surface Chemistry in Mineral Processing', *Chem. Ind. (London)*, 1975, 54.

M. C. Bennett, 'A New Industrial Process for Decolorising Sugar', *Chem. Ind. (London)*, 1974, 886.

D. Attwood and A. T. Florence, 'Physico-Chemical Principles of Pharmacy', Macmillan Press, London, 1981.

'Drug Delivery Systems', News Item, *Chem. Ind. (London)*, 1985, 6.

S. J. Douglas and S. S. Davis, 'The Use of Nanoparticles in Drug Targetting', *Chem. Ind. (London)*, 1985, 748.

R. A. Dawe and E. O. Egbogah, 'The Recovery of Oil from Petroleum Reservoirs', *Contemp. Phys.*, 1978, **19**, 355.

C. W. Arnold, 'Enhanced Oil Recovery', *Chem. Technol.*, 1977, **7**, 762.

T. J. Jones, E. L. Neustadter, and K. P. Whittingham, 'Water-in-Crude-Oil Emulsion Stability and Emulsion Destabilisation by Chemical Demulsifiers', *J. Can. Pet. Technol.*, 1978, **17**, (2), 1.

D. E. Graham, A. Stockwell, and D. G. Thompson, 'Chemical Demulsification of Produced Crude Oil Emulsions', in 'Chemicals in the Oil Industry', ed. P. H. Ogden, Special Publication No. 45, The Royal Society of Chemistry, London, 1983, p. 73.

J. F. Marsh, 'Colloidal Lubricant Additives', *Chem. Ind. (London)*, 1987, 470.

Symbols and Terminology

The following publications summarise the I.U.P.A.C. recommendations concerning colloid and surface chemistry:

'Manual of Symbols and Terminology for Physicochemical Quantities and Units':

Appendix II: 'Definitions, Terminology and Symbols in Colloid and Surface Chemistry';

Part I: *Pure Appl. Chem.*, 1972, **31**, 579—638;

Part I.13: 'Definitions, Terminology and Symbols for Rheological Properties', *Pure Appl. Chem.*, 1979 **51**, 1213—1218;

Part I.14: 'Light Scattering', *Pure Appl. Chem.*, 1983, **55**, 931—941;

Part II: 'Heterogenous Catalysis', *Pure Appl. Chem.*, 1976, **46** 73—90.

Appendix III: 'Electrochemical Nomenclature', *Pure Appl. Chem.*, 1974, **37**, 503—512.

'Reporting Experimental Data Dealing with Critical Micellisation Concentrations', *Pure Appl. Chem.*, 1979, **51**, 1083—1089.

'Reporting Physisorption Data for Gas/Solid Systems', *Pure Appl. Chem.*, 1985, **57**, 603—619.

'Reporting Experimental Pressure–Area Data with Film Balances', *Pure Appl. Chem.*, 1985, **57**, 621—632.

'Reporting Data on Adsorption from Solution at the Solid/Solution Interface', *Pure Appl. Chem.*, 1986, **58**, 967—984.

See also:

I. Mills, T. Cvitas, K. Homann, N. Kallay, and K. Kuchitsu, 'Quantities, Units and Symbols in Physical Chemistry', Blackwell Scientific Publications, Oxford, London, *etc.*, 1988.

Subject Index